THE HUMMINGBIRD HANDBOOK

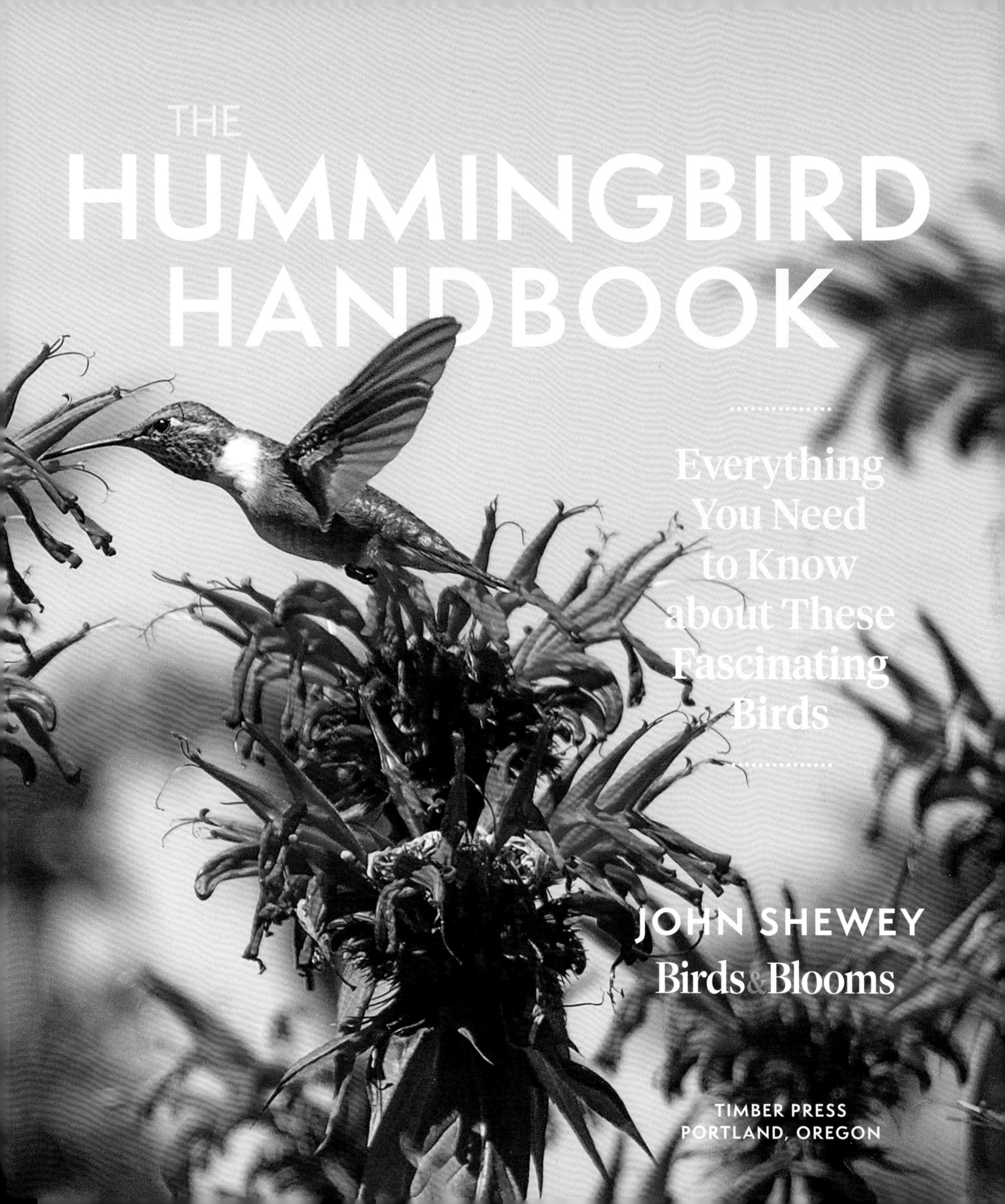

THE HUMMINGBIRD HANDBOOK

Everything You Need to Know about These Fascinating Birds

JOHN SHEWEY
Birds & Blooms

TIMBER PRESS
PORTLAND, OREGON

Range maps courtesy the Cornell Lab of Ornithology.
Photo credits appear on page 230.

Published in 2021 by Timber Press, Inc.

The Haseltine Building
133 S.W. Second Avenue, Suite 450
Portland, Oregon 97204-3527
timberpress.com

Printed in China

Text and cover design by Mary Winkelman Velgos
ISBN 978-1-64326-018-1

Catalog records for this book are available from the Library of Congress
and the British Library.

In the end, we will conserve only what we love;
we love only what we understand, and
we will understand only what we are taught.

—BABA DIOUM, 1968

CONTENTS

FOREWORD

Like the whir of their teensy wings, hummingbirds create quite a buzz wherever they show up. Seeing one of these exuberant and energetic fliers dash through your garden, hover at a flower you planted or sip from your feeder for the 100th time is just as exciting as the first time. You're not alone if your heart swells just a tiny bit when a hummingbird chooses to make a stop in your yard—it happens to me, too.

If there's one thing I've learned in my 10+ years with *Birds & Blooms*, it's how much people love and adore hummingbirds. I see it in the high engagement on social media posts involving hummingbirds. I see it in the thousands of hummingbird photos submitted to our photo contests. I see it in the positive reactions to our annual hummingbird edition of the magazine. Our audience yearns for more photos and information about the tiny birds that spark joy and wonder with every wingbeat.

At *Birds & Blooms*, we tend to focus on hummingbirds in the backyard. But in this book, John Shewey offers up everything you've ever wanted to know about hummingbirds, so teaming up on this project made perfect sense. He captures the spirit and allure of these captivating birds in every fascinating fact, historical tidbit, amusing anecdote, species profile and plant pick. It's truly a handbook—a complete guide to the lives, movements and habits of hummingbirds. Here, he gives you all of the tools you need to understand what makes them tick, how to attract them to your backyard with feeders or plants and all of the basics to create a bustling hummingbird habitat.

If you're like me and call the eastern half of the U.S. home, we only experience ruby-throated hummingbirds whizzing around our yards. Those living in the western U.S. have opportunities to see many more species. I especially appreciate John's thorough reporting on how to locate hummingbirds if you're traveling within North America, Central America or South America. He includes detailed information about festivals dedicated to hummingbirds, hot spots of the western U.S. and how hummingbirds contribute to ecotourism outside of North America.

At any location known for hummingbird sightings, even if it's your local park, you're sure to find fellow bird enthusiasts just like you. Because we're a community, we hummingbird lovers, and John and I are glad you're here!

Kirsten Schrader
Executive Editor, *Birds & Blooms*

The Allure of Hummingbirds

As I write this, on a late-summer day with the windows open, my yard is a cacophony of hummingbird squeals and chatters. The tiny birds are chasing one another nonstop, each trying to defend its food sources. I often wonder how they ever manage to find time to actually feed themselves—on flower nectar, tiny bugs, and the sugar water in my feeders. How do they replenish their rapidly expended energy when they are so obsessed with chasing each other away? And then I realize I have lost track of time and whiled away 30 minutes just watching these beguiling little winged acrobats.

The approximately 340 extant hummingbird species are entirely and uniquely American, living from Alaska to southernmost South America. Superlatives spanning the dictionary are insufficient to describe them. These captivating birds mystified early explorers, baffled scientists, enraptured ornithologists—and they continue to do so, as researchers routinely make intriguing discoveries about hummingbirds. Unsurprisingly, hummingbirds enthralled indigenous peoples; many cultures throughout the Americas included these bejeweled

Ruby-throated Hummingbird.

creatures in their mythologies. And when European settlers arrived in the New World, they found hummingbirds so enigmatic that many people were frightened by them. Hummingbirds were unlike anything they had ever seen in the Old World.

These perplexing and enchanting creatures mesmerize us with their unparalleled combination of dazzling psychedelic colors and incredible speed, all packed into diminutive bodies. Life for a hummingbird is a blur to us; in fact, their exact modes of hovering and flying and feeding were only recently unraveled: it took advanced high-speed motion photography, and the subsequent slowing of hummingbirds on film, for their actions to be deciphered.

I regard hummingbirds as portals into the natural world. They are universally fascinating and appealing—so much so that they are empowered to draw people into a deeper appreciation and understanding of ecology. They foster curiosity among observers, and with luck such curiosity will then transcend the messenger—the hummingbird—to engender a rich interest in natural sciences. My own interest in hummingbirds began in my childhood when my mother put out feeders, but my fascination with them commenced on 1 June 2012.

On that early-summer day, I was immersed in a five-day bird-photography expedition with Tim Blount for our book, *Birds of the Pacific Northwest*. Atop a sprawling fault-block mountain in remote southeastern Oregon, we stopped at an outpost of the U.S. Fish and Wildlife Service, something of an oasis in the high desert thanks to trees planted there decades ago. They attract migrating songbirds, but we hardly expected the whir of hummingbird wings when we stepped out of the truck. Blooming shrubbery amid a vast sea of sagebrush had obviously attracted these hummingbirds, which were refueling during the peak of migration, when northern species travel many hundreds of miles between their wintering grounds in Mexico and their breeding territories.

Armed with cameras and binoculars, Tim and I assumed the handful of hummingbirds to be the common local species—Rufous or Black-chinned Hummingbirds—so we were astonished to discover that, instead, they were Broad-tailed Hummingbirds, very rare in Oregon. Two males were competing for the attentions of two females, and all four were buzzing about at warp speed, alternating between chasing one another and lapping nectar from the abundant blooms. We captured many fine images, and Tim eventually walked off into the surrounding steppe in hopes of photographing other birds. I stayed behind with the hummingbirds and I'm glad I did, because when I walked around to the far side of the row of flowering shrubs, I discovered a fifth member of the tribe—but this one was different. I soon identified it as a male Ruby-throated Hummingbird.

The Ruby-throated Hummingbird is the common species of the entire eastern half of the continent. Only a widely straying individual—a lost bird, essentially—would show up in Oregon, and indeed this turned out to be only the third

Ruby-throated Hummingbird sighting ever recorded for the state. I pondered his unlikely journey, marveled at his ability to survive the sub-freezing nights along the way, far, far away from where he was hatched and fledged, and where his species thrives. Further, Tim and I were astounded that we had found both these species, one very rare in Oregon and one virtually unheard-of in Oregon, at the same place and time. How they got there, why they ended up there, where they would go from there—we contemplated these mysteries, and thus began my fascination with these marvels, the hummingbirds.

They are tiny, tough, and adaptable; they live at a frenzied pace; hummingbirds elicit awe and wonder. I can certainly relate to German pastor Gottlieb Mittelberger (1714–1758), who, upon seeing a hummingbird in Pennsylvania, declared it the "most wonderful bird . . . perhaps in the whole world." He went on: "This little bird . . . glitters like gold, and sometimes it appears green, blue and red. Its beak is rather long, and as sharp as a needle; its feet are like fine wire. It sips only honey from flowers; hence it has the name of a sugar-bird [and] moves its wings with indescribable swiftness."

These most wonderful birds are full of surprises. On a brumal November morning, I was waiting for an airport shuttle at a deserted parking lot near downtown Portland, Oregon. Leaden clouds hung over gray concrete buildings; an icy wind pushed tendrils of hoary fog through dreary alleys; the occasional pedestrian shuffled by, huddled against the chill. But then I was startled by a sudden blur nearly at my feet: a spritely little Anna's Hummingbird buzzed to a hovering stop at a tiny upstart shrub, fed briefly from the meager remaining blooms, moved on to the next shrub, and then darted off, up and away, as I traced her path out of sight, my gaze uplifted along with my spirits.

ONE

Hummingbird Trivia: Facts, Fictions & Folklore

Hummingbirds have fascinated humans for millennia. Long before European explorers set foot in the New World—the only place on Earth where hummingbirds live—native peoples throughout the Americas were enraptured by hummingbirds; some cultures revered them and even used hummingbird feathers and skins in their clothing. When Europeans reached the Americas on the eve of the Age of Reason, they too were smitten by hummingbirds, and so across world cultures, myths, legends, and fanciful fictions surrounded these tiny birds.

Most early explorations of the New World included naturalists either by definition or by hobby. They encountered a natural and cultural world completely foreign to them, and lands populated by creatures never seen outside of the Americas. As specimens poured into Europe, the scientists of the day began to unravel some of the mysteries surrounding hummingbirds, but even to modern science, hummers remain enigmatic in many ways, and ongoing research continues to fill holes in our understanding of these amazing little birds.

Anna's Hummingbird.

With hummingbirds, the facts often prove more fascinating than the fictions, and as researchers unlock the secrets about the ecology and biology of these birds in all their diversity, more questions arise. That of course, is the nature and beauty of science: questions beget answers through application of scientific method, and answers in turn almost always beget further questions. This chapter delves into those questions and answers and more questions about these inscrutable birds.

ABOVE A close-up of the head of a Costa's Hummingbird shows the specially structured feathers that help create iridescent colors.

BELOW With a turn of its head, a male Rufous Hummingbird suddenly and startlingly reveals a glittering orange gorget.

All That Glitters

With just a turn of the head, a hummingbird explodes in iridescent radiance, its gorget—the patch of colorful feathers covering its throat—instantly conflagrant in shades that cover the color spectrum, depending on the species. These dazzling colors derive from the feather structure rather than from pigmentation. Each tiny iridescent feather has barbules that are densely packed with layers of microscopic platelets filled with air bubbles. These platelets reflect and refract light waves, explains Bob Sundstrom of *BirdNote*, "creating color in the manner of sun glinting off oily film on water."

The color you see in a hummingbird's gorget depends on the angle and quality of the light being reflected and refracted by the specialized throat feathers (aka spangles). In one moment the gorget (and in some species, the crown as well) appears black, deep gray, dark bronze, or deep purple, and then the bird turns its head and the throat bursts into magnificent colors like a neon flashbulb. Iridescent gorgets and crowns are most prevalent in males, with perhaps 70 percent of species being sexually dimorphic (that is, males and females look different). In some dimorphic species, even the females have at least a partial gorget of iridescent feathers.

Hummers vary tremendously in the color and arrangement of their iridescent parts. Even among the handful of hummingbird species that are widespread in the United States, the variety in colors is impressive, from the blazing orange gorgets of male Allen's and Rufous Hummers, to the incredible iridescent magenta throat, face, and crown of the beautiful male Anna's Hummer, to the closely related Costa's Hummingbird, whose head is enwrapped in royal purple, with a gorget that tapers into long mustachelike points on each side of his head.

The term "gorget," incidentally, derives from the shaped metal sheet worn as part of body armor beginning in the Middle Ages. By the 1500s, the plate-metal gorget had reached its zenith in both design and function, and soon became ornamental as well, signifying status and rank, a tradition that continues today in some militaries. It seems an apt moniker for the vibrant adornment of hummingbirds—and yes, the word "gorgeous" is etymologically related!

The vast majority of hummingbird species live in Mexico, the Caribbean, Central America, and South America, and in these more southerly latitudes, hummers reach the pinnacle of their mind-boggling diversity. Some species sport iridescent feathers over nearly their entire bodies; others combine incredible iridescence with opulent adornments, such as streaming bannerlike tails and flamboyant crests.

ABOVE This male Violet-capped Woodnymph (a South American species) is extensively adorned in iridescent colors.

BELOW As this peacock demonstrates, hummingbirds have no monopoly on iridescence, but they do take the concept to remarkable extremes.

The amazing hummingbirds, however, have not cornered the market on iridescent feathers. Many other birds share this trait, including grackles and blackbirds, swallows, various ducks, and peacocks. But few would argue that hummingbirds take iridescence to magnificent extremes.

Extremes Among the Extreme

All hummingbirds range from small to tiny—almost.

The largest of the 340-odd species is—you guessed it—the Giant Hummingbird (*Patagona gigas*). Imagine a hummingbird 9 inches long with a wingspan of more than 8 inches. This enigmatic denizen of the Andes is bigger than a Barn Swallow and as long as a Robin, though much slenderer. And like all hummers, even this behemoth is extremely light in weight, at about 0.85 ounces, and capable of amazingly acrobatic flight; but compared to much smaller hummers it appears somewhat cumbersome and hovers at a rate of only about 15 wingbeats per second—on the slow side for a hummingbird.

At the other end of the hummingbird size spectrum, myriad species weigh in at a mere fraction of the bulk of the Giant Hummer. But even among the Lilliputians, there is a standard bearer: the tiny Bee Hummingbird (*Mellisuga helenae*), endemic to the Cuban archipelago. The world's smallest bird of all, the Bee Hummingbird weighs less than a tenth of an ounce—less than a U.S. penny—and is just barely over 2 inches long.

ABOVE **The Giant Hummingbird, widely distributed in the Andes, is by far the largest species of hummingbird.**

BELOW **The minuscule Bee Hummingbird of Cuba is the world's smallest bird.**

Fit the Bill

A casual observation might suggest that hummingbirds use their long, thin, dainty beaks like straws; however, beginning early in the 19th century, scientists realized that the tip of a hummingbird's tongue forks into two tiny tubes, and so they postulated that the birds must drink nectar through capillary action, the same mechanism that allows a towel to draw in water. Turns out, the scientists were wrong—for well more than a century.

Only in recent years have scientists figured out how a hummingbird laps up nectar with its long, slender tongue.

Inquisitive scientists and high-speed motion photography finally cracked the code in 2011, when researchers Margaret Rubega and Alejandro Rico-Guevara discovered that hummingbirds feed via a pistonlike method. They lap up nectar with their tongues, the tiny forks at the tip springing open to gather fluid; then the tongue retracts as the bill squeezes shut, compressing the tongue and allowing the bird to lap up the nectar. They repeat this high-speed lapping 15 to 20 times per second.

Another enduring mystery—among the many with these enigmatic birds—is how they catch insects, which make up a significant and important part of their diets. Again, relying on high-speed frame-by-frame photography, researchers learned that hummers can flex their lower bill downward to get it out of the way and widen the base, and then snap the bill closed at blinding speed. Combined with their aerial agility, this adaptation for catching insects on the wing allows hummingbirds to obtain life-sustaining protein, fat, amino acids, and other important nutrients.

Bill lengths and shapes vary dramatically throughout the hummingbird world, and some species coevolved with specific flowers that provide their primary nectar sources. Many hummingbirds have bills specifically adapted to fit certain flower species, and both the shape and length of the bill, and the shape and structure of the flower tube can coevolve regionally and temporally between two species. Studies in the evolutionary relationship between pollinators, such as hummingbirds and bees, and flowers that need to be pollinated,

LEFT The Sword-billed Hummingbird has by far the longest bill relative to body length of any bird in the world.

RIGHT Northern banana passionflower coevolved with the Sword-billed Hummingbird, its primary pollinator.

continue to provide amazing insights. One such research project, headed by Lena Hileman at the University of Kansas, revealed that the flowers of various *Penstemon* species show either bee or hummingbird adaptation: species that are adapted to bee pollination are generally bluish or purplish, with a flower tube of sufficient diameter to allow bees to enter and a stamen positioned to deposit pollen on the backs of bees. Conversely, species adapted for pollination by hummingbirds are red or orange-red with narrow openings to allow only the bill and/or tongue of a hummingbird to enter; they don't need to offer a landing pad.

All this coevolution has led to some truly bizarre flowers but also to some hummingbirds with extraordinary bills. Among the most idiosyncratic is the Sword-billed Hummingbird, native to Andean South America. This bird's daggerlike bill is nearly as long as its entire body, stretching to almost 4 inches. This long bill allows the Sword-billed Hummer to feed on long-tubed flowers whose nectar is inaccessible to other species, especially and most notably the beautiful pink blooms of northern banana passionflower (*Passiflora mixta*). The dramatic coevolution of these two species, which occupy the same range in the Andes, is an example of a mutualistic relationship: the flower is pollinated by the hummingbird, and the bird is rewarded with a rich source of nectar.

American Idols

Hummingbirds are as American as apple pie; or more accurately, as American as *pastel de manzana* and *torta de maçã* (Spanish and Portuguese, respectively, for "apple pie"). After all, hummingbirds are native only to the Americas, mostly Central and South America. In Brazil,

RIGHT Huitzilopochtli, the hummingbird war god, was central to Aztec mythology.

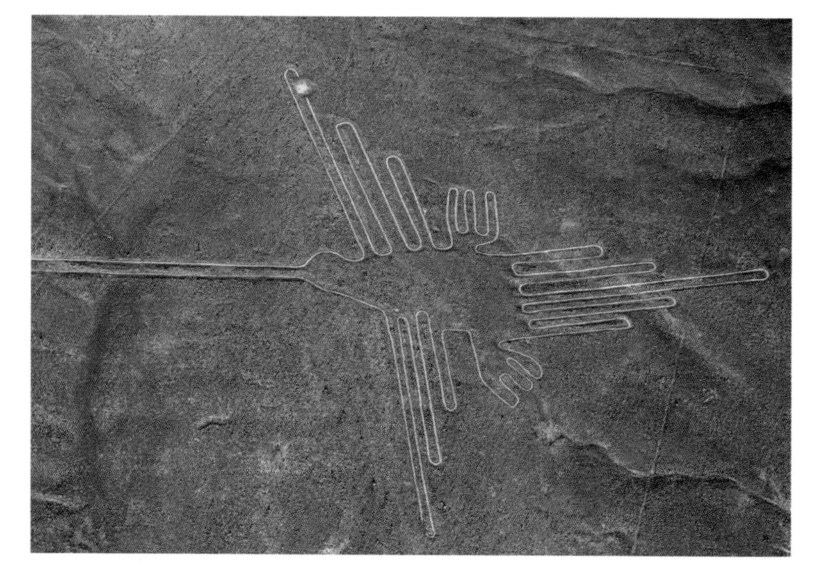

BELOW The famous Nazca Lines in Peru include numerous animal figures, including this hummingbird.

a Portuguese-speaking nation, the elegant word for hummingbird is *beija flor* (from *beijar*, "to kiss," and *flor*, "flower")—hence we arrive at "flower kisser," more or less—certainly a visually descriptive moniker. In Spanish, the predominant language of Central and South America, the word for hummingbird is *colibrí*, which is thought to derive from the French Caribbean.

This intriguing Arizona petroglyph depicts a hummingbird.

In Latin America, colloquial names for hummingbird are often more colorful. Mexican synonyms include *chuparosa* ("rose sipper"), *florimulgo* ("flower milker"), *pájaro mosca* ("fly bird"), *chupamiel* ("honey sipper"), *joyas voladoras* ("flying jewels"), and others. Indigenous peoples of the Americas had their own names for these tiny birds; to the Mayans, they were *x-ts'unu'um*.

Hummingbirds are key players in Native American mythology, and perhaps nowhere is this more evident than in the massive hummingbird mosaic in the Nazca Lines in the Nazca Desert of southern Peru. This geoglyph, 320 feet long and 216 feet wide, was created around 2,000 years ago by the indigenous Nazca people by digging through the topsoil layer of dark pebbles to reveal a lighter substrate. Farther north, in what is now Central America and Mexico, hummingbirds played a significant role in Mayan and Aztec culture. The Aztec hummingbird deity, Huitzilopochtli, is a god of war to whom they made human sacrifices; other Native American cultures attributed special significance to hummingbirds, which are central characters in many indigenous legends.

Nearly as soon as they set anchor in American waters, European explorers became enamored of hummingbirds. As was the customary scientific process, naturalists of the era collected as many specimens as they could, and voluminous collections accrued in the hands of European naturalists, artists, museums, universities, and, notably, private individuals of considerable means. In some cases, these preserved hummingbirds served as models for beautiful artistic renditions, often included in books about New World birds.

But the European discovery of America's hummingbirds also coincided with the Continent's craze for feather-adorned fashions, and that calamitous convergence proved disastrous for countless avian species.

Trouble in Paradise

Unfortunately for the hummingbirds, Europeans soon began desiring them to make expensive clothing, jewelry, flowerlike arrangements, and other art pieces. Not long ago, a British auction house offered for sale—for about $50,000—an elaborate Victorian-era fireplace with a diorama of more than 100 stuffed hummingbirds arranged on branches mounted between two large panes of glass and framed in gilded, delicately carved metal. The 19th-century world of high fashion fostered a fascination with colorful and exotic creatures from the New World, Asia, and beyond; and the demand for fancy-feathered women's hats drove a British, Continental, and even American craze for brilliant plumes, feathers, and even entire stuffed birds, an infatuation that decimated countless bird populations throughout the world.

Talk about suffering for fashion: hummingbirds—and many other species—were killed by the millions in the late 1800s, their skins and feathers destined for the millinery trade.

Luckily, however, by the early 20th century, outcry led to reform, first in the United States, later in England. Though George Bird Grinnell, the influential editor of *Forest and Stream*, was perhaps the first to publicly rail against the absurd excesses of the feather trade in the 1880s, among the most vocal and insightful of the change-minded crusaders aiming to eliminate the merciless destruction of bird populations was William Temple Hornaday, whose book *Our Vanishing Wildlife* (1913) brought needed public and political attention to the plumage trade that was rapidly destroying whole bird populations. James Buckland, the tireless champion of the English plumage bill that would eventually curtail that country's participation in the decimation of the world's birds, ran the shocking numbers in December 1913: "Here are the totals of just a few species whose plumage has been sold during the past twelve months at the London feather sales: 216,603 Kingfishers; 21,318 Crowned Pigeons, 20,715 quills of the White Crane; 17,711 Birds-of-Paradise; 5,794 pairs of Macaw wings; 4,112 Hummingbirds; and so on, through the whole list of brilliantly plumaged birds."

Hummingbirds in Central and South America were ruthlessly hunted to sate the demands of high fashion—in one weeklong London auction, a staggering 12,000 hummingbirds were among an unfathomable 350,000 bird skins sold. Hummingbirds were prized for their magnificent iridescence, and in 1914, American journalist Rene Bache wrote a scathing indictment, carrying the sardonic headline, "By All Means Buy A Humming Bird Coat, They are the latest thing and the cost is only a trifling $25,000." That's about $650,000 today. Bache decried that the knee-length coat required thousands of hummingbird skins, but the insatiable demand in Europe to create ever-more elaborate fashions meant entire species were pushed to the brink of extinction.

Thankfully public outrage was growing sufficiently that new and revolutionary wildlife-protection laws were being debated and enacted. In 1914, *Bird Lore*, the bimonthly magazine published by the fledgling Audubon Society, reported thus:

> In Paris, France, on March 9, 1914, the woman's paper, *La Vie Feminine*, gave its inauguration reception at the Galerie d' Excelsior, 88 Champs Elysees. The lecturer was the novelist Pierre Loti (Lieut. L. M. J. Viaud), who was asked to speak about women in Turkey. His opening remarks may be of interest to the Audubon Society: "Looking at you from this platform, I see a surging mass of feathers, which your hat-makers insist upon placing—some straight in front, others over one ear, then again a plume trails over the back of the head, in a weeping-willow style, giving the impression of unrest. I will end my digression by telling you something profoundly sad. Among the plumes on your hats I distinguish innumerable [egrets], quantities of Birds-of-Paradise, and, as I turn my eyes away, I think of the ruthless massacres which bird-hunters are carrying on for your pleasure and vanity. Poor little winged world, inoffensive and charming, which in half a century, thanks to you, will be found nowhere! I recall some specimens, the most wonderful, which have already disappeared, with no possible return. What a sacrilege! What a crime! To have sent into oblivion a species of bird-life which no mortal can re-create in this world! Ladies, I ask mercy for the birds of fair plumage. Believe me, all of you will be just as lovely, and appear less cruel, when you have discarded the covering of these little bodies, which you now wear on your hats."

Hummingbirds and many other members of the avian world were spared, as strict laws were enacted to protect them, and hummingbirds are now celebrated throughout the Americas. These feathered gems even support a robust subculture in ecotourism. Hummingbird fans flock to locations known to support diverse and intriguing species. Throughout the Americas, tour guides, lodges, parks, sanctuaries, festivals, and even entire communities cater to and are supported by itinerant enthusiasts who love these little American icons.

ABOVE LEFT Many nations in the Caribbean, Central America, and South America have featured hummingbirds on currency—for example, this Suriname banknote, depicting a Long-tailed Hermit.

ABOVE RIGHT The Green-throated Carib, at rest, on a beautiful postage stamp from Montserrat, in the Lesser Antilles.

We're on the Money!

And on the stamps, and even on the coins. Hummingbirds are so universally beloved that they have appeared on currency, coins, and postage stamps throughout the Americas, from Brazil to Canada. In many instances, currency and stamps depicting hummingbirds present beautifully rendered artwork, and collectors could easily specialize in collecting postage stamps or currency adorned with these mesmerizing birds. Several times, the United States Postal Service (USPS) has issued hummingbird stamps. In 1992, for example, the USPS issued a set of five stamps depicting five different North American species: the Ruby-throated, Broad-billed, Costa's, Calliope, and Rufous Hummingbirds. Most Central and South American nations, including Caribbean nations, have issued hummingbird postage stamps as well, but they alone—rather than the United States or Canada—have also, in many cases, given hummingbirds a prominent place on currency.

Spread the Wealth

Hummingbirds and the flowers they feed from help one another, a textbook example of a symbiotic relationship. The birds seek sugary nectar from within the blooms, and in so doing, inadvertently gather pollen, which sticks to their bills and forehead. When the birds feed at other flowers of the same species, the pollen is transferred from plant to plant. Bees and butterflies also cross-pollinate flowers, but through evolutionary processes, countless flowering plants have developed characteristics that favor pollination by one type of animal or another. Flowers evolved for hummingbird pollination tend toward bright shades, lack strong scent (insects use scent to find flowers; birds use vision), and are structured in ways that allow hummingbird bills and tongues to reach the nectar while excluding bees and butterflies from doing so. Throughout the Americas, in fact, many species of flowers rely exclusively on hummingbirds for pollination. For example, in eastern North America, fire pink (*Silene virginica*), a bright red wildflower, is primarily pollinated by the familiar Ruby-throated Hummingbird—just one of countless examples of a species dependent on hummingbirds.

ABOVE The powdery yellow substance on this hummingbird's bill is flower pollen; with luck (from the plant's perspective), she'll soon visit another plant of the same species and cross-pollinating will be achieved.

BELOW Hummingbirds can maintain their aerial balance in wind, adjusting as needed to keep pace with a flower that is dancing in the breeze.

Aerial Acrobats Beyond Compare

The bird world is full of amazing flyers: falcons diving at death-defying speeds, swifts and swallows maneuvering in hot pursuit of bugs, accipiters dodging through dense forest at perilous velocity. But no bird can match the hummingbird for agility on the wing. Hummingbirds are experts at hovering, the strategy employed by most species for feeding on the nectar of flowers. They can also fly backward and even, for brief moments, upside down. Some species perform amazing courtship flights, diving from aloft at bedazzling speeds only to check their plunge suddenly, tracing an arc too fast for the human eye to track.

Hummingbirds have evolved an insectlike flight adaptation: other birds create the lift needed for flight with each downstroke of the wing, but on the upstroke, the outer wing folds inward toward the body in preparation for the next downstroke. In other words, typical bird flight is achieved by flapping the wings up and down. Hummingbirds, however, rotate, or twist, their upper arm bones (humeri) to invert the wing and gain lift from the upstroke as well as the downstroke. The result? The most agile birds on the planet.

Now That's a Blur!

The advent of high-speed motion photography has allowed scientists to accurately measure and study wing movements that are way too fast for the human eye to see. We now know that, depending on the species, hummingbirds flap their wings from about 10 to more than 80 times per second while hovering, and even faster during the courtship display dives of some species.

Moreover, hummingbirds can fly very fast, routinely attaining speeds of 20 to 30 miles per hour and triple that during the amazing courtship display dives used by males of some species. The swiftest hummingbirds are extremely fast: Christopher James Clark, who has done extensive hummingbird research at the UC Berkeley, discovered that during its courtship dive, the male Anna's Hummer folds its wings in to its sides, achieving an average maximum velocity of approximately 90 feet per second, the highest speed every recorded for a vertebrate, relative to size. Even more incredibly, as it pulls up, with wings spread, the bird experiences centripetal accelerations nearly nine times greater than gravitational acceleration. In short, hummers are faster than a fighter jet, relative to size, and withstand g-forces that would make the average person black out.

ABOVE This female Rufous Hummingbird's wings beat too fast for the human eye to track.

BELOW Dueling Rufous Hummingbirds in Arizona.

Pint-size Pugilists

Many hummingbird species behave aggressively toward other hummers. Even females and juveniles will defend their favorite feeding locations, but such assertiveness is especially pronounced with territorial males. A dominant male will defend "his" favorite flowers or feeder ruthlessly and relentlessly, chasing away all male interlopers. He may even chase away females, though often perceived aggression toward females of his species may be related to courting behavior. Some male hummingbirds routinely harass females until the female breeds with him, at which point she typically becomes part of the territory; however, whether or not a female builds her nest within the territory of the male with whom she mated is an open question.

Male hummingbirds defending their breeding or feeding territories typically tolerate no intrusion from fellow males of their own species and will even chase away males and females of other species. The consensus is that breeding territories may cover up to about a quarter of an acre, perhaps more, especially in open country. Feeding territories seem to vary considerably in size and scope, at least based on my own observations with Anna's, Rufous, and Black-chinned Hummingbirds. A hummingbird's feeding territories may include not only hummingbird feeders but also particular flowers, flowers beds, gardens, water features, or entire yards. Certain species are more bellicose than others; in one instance I had a female Rufous Hummer that so truculently guarded the feeder she claimed that she gave no quarter to either sex, of any species.

Territorially aggressive behavior in hummingbirds takes several forms, and the interactions, rarely damaging to the birds, are frenetic, fast-paced, and fascinating. It may begin with excited scolding and brilliant displays of the gorget, but the classic territorial encounter is the aggressive chase in which one hummer drives the interloper away at blinding speeds, usually chattering loudly. Rarely a brief physical battle can ensue, and in those cases, hummers can injure or even kill one another with their bills. This behavior can easily be confused with breeding-season aggression in which males actively chase females, especially when both sexes are freshly arrived from their spring migration. Once things settle down a bit, courtship season begins in earnest, and that's when the males perform their amazing, acrobatic courtship display flights aimed at duly impressing the observing female.

During the nonbreeding season, hummingbirds may be less aggressive about defending feeding territories, but they'll still try to monopolize a hummingbird feeder. Based on my observations with Anna's Hummingbirds, sometimes more than one male may routinely use the same feeder, even aggressively defending it, only to have another male replace him in that role at different times during the day.

Moreover, even during the breeding season, dusk can bring a truce to the world of warring hummingbirds. The birds tend to feed heavily at and just after sunset and, with bellies full, they become impressively tolerant of one another, even with regards to species often considered the most territorial, such as Anna's and Rufous. On the other hand, I once had a particularly pugnacious male Rufous who, as the dusk feeding time arrived, made it his personal mission to prevent all interlopers from getting within about 30 feet of his feeder. And when another male Rufous tried to violate his space, bantamweight fisticuffs would ensue—physical fights between hummers happen very quickly, but I had a front row seat. During the gloaming, when that little miscreant was driving every other hummer away from his feeder, I would often peek around the other end of the house, where another feeder would attract as many as six hummers of two different species, all fueling up in the typical harmony for the night to come.

In another example of sunset accord, years ago a restaurant in Aspen, Colorado, had hung hummingbird feeders from the eaves, just outside the windows, along two sides of the building, more than a dozen feeders in total. Seated for dinner while the summer sun still glowed brightly to the west, I watched two male Broad-tailed Hummingbirds defend the entire collection of feeders from interlopers—one bird held sway on the east side of the restaurant, the other dominated the south face. They were tolerant of a few females coming to feed but didn't extend the same courtesy to males. But a remarkable transformation occurred just as the sun set: dozens of hummingbirds, males and females, descended on the feeders and peacefully fed alongside one another.

Pick on Someone Your Own Size!

Many hummingbirds don't limit their belligerence to members of their own tribe. They will aggressively intimidate and chase off other birds and even mammals. I've witnessed Rufous Hummers attempt to drive off various larger birds and once watched a hummer buzzing at full tilt amid a flock of geese in flight, harassing one after another. They will even take on birds of prey ranging from tiny falcons to eagles; many observers have seen hummers harassing accipiters—Sharp-shinned Hawks, Cooper's Hawks, Northern Goshawks—which are specialists in the art of killing and eating other birds, though hummers are generally too small for them to bother with.

Some years ago, fascinated, I watched a male Rufous Hummingbird, among the most truculent of all species, go out of its way to harass a Song Sparrow that had done nothing to deserve such badgering: the hummingbird had been routinely feeding at a side-yard feeder adjacent to my birdseed feeders, set up for sparrows, grosbeaks, finches, and the like. Many hundreds of times that summer month, the hummer had gone about his feeding unperturbed by the activity nearby. Then one day while I watched out the window, he suddenly decided a random Song Sparrow, one of several, that was sitting on a platform minding its own business, would be the object of his bullying. The hummer zipped over, hovered a foot above the sparrow, uttered aggressive squeaks, flared its gorget and tail, and dove toward the sparrow three times in succession. The sparrow cowered; the hummer, apparently satisfied with the effects of his tyranny, flew off to the other side of the house—probably in search of other victims. Two weeks later I watched that same bellicose hummer chase a Song Sparrow 100 feet out into the chest-high grass in the nearby bottomland, and when the sparrow ducked into the grass for cover, the hummer dove in and out three times until satisfied his victim was out of reach. Three days later I watched him chase a Black-headed Grosbeak for 50 yards.

In an episode reminiscent of my hummer-versus-geese incident, Philip Henry Gosse, in *The*

A staredown between a Red-tailed Hawk and a Broad-tailed Hummingbird.

Birds of Jamaica (1847), recounted, "I have been amused to see a Hummingbird chasing a Heron; the minuteness and arrowy swiftness of the one contrasting strangely with the expanse of wing and unwieldy motion of the other. The little aggressor appears to restrain his powers in order to annoy his adversary, dodging around him and pecking at him like one of the small frigates of Drake or Frobisher peppering one of the unwieldy galleons of the ill-fated Armada."

Life on the Edge

Hummingbirds live life at a pace we can't comprehend. But their hyper-speed existence carries a tremendous cost, a burden they must bear. Hummingbirds consume energy rapidly and efficiently to maintain their no-holds-barred frenzy of feeding and fighting and flattering. And when environmental demands threaten their very existence, they have a built-in defense mechanism called torpor, a kind of semi-hibernation they enter primarily to survive cold nights when they are not out feeding—refueling—countless times as they do throughout the daylight hours.

During the night, hummingbirds seek secluded perches, usually deep in cover. They lower their metabolism by as much as 95 percent, slowing their heartbeat to around 50 beats per minute from their active rate of hundreds of beats per minute (a Blue-throated Mountain-gem's heart rate was once measured at 1,260 beats per minute). Their breathing slows, and they enter a state of hypothermia in which their body temperature drops to a set point, a temperature just barely adequate to sustain life. In the hours before dawn, torpid hummingbirds begin to awaken, shivering to gain body heat; they need about 20 minutes to fully recover from torpor, and then they are off to find the first of many meals of the day.

Fluffed up to preserve heat, this little Costa's Hummingbird is still asleep early in the morning.

This extreme lifestyle—a life on the edge of death, essentially—requires such constant refueling that I often wonder what the mortality rate might be for torpid hummers—how many never awaken at all?

All that aerial agility and the warp-speed hummingbird lifestyle exacts a heavy toll: hummingbirds use a lot of energy and use it very quickly. Consequently the most time-consuming activity in the everyday life of a hummingbird is sitting on a favorite perch between bouts of feeding and freneticism. Their frequent rest breaks, which allow them to digest sugars gathered from flowers and feeders, account for 75 to 80 percent of their time. During periods of peak activity, such as when males are performing courtship displays or when females are building nests and feeding young, hummers often feed at regular short intervals, such as every 8 or 10 minutes.

You're on Your Own, Honey

While male hummingbirds get all the glory, females do all the work: males of the majority of species are flamboyantly colored, but the female hummer, usually duller in color, builds the nest, sits on the eggs, and raises the young all by herself. It's a monumental task, of course, but one she is genetically programmed to carry out.

The nest-building process is as arduous as it is fascinating. Most hummingbird species build tiny cup-shaped nests from downy plant materials, such as the ultrasoft seed pods of thistles, cattails, and cottonwoods, along with tiny twigs and other organic materials. Amazingly, they bind these materials together and anchor the nest to a branch using spiderweb filaments. Most species camouflage the outside of the nest with tiny bits of bark, lichen, and leaf. Frequently the nest is attached to a tree or shrub limb and is so small it easily escapes notice, looking much like a small knot in the wood. Hermit hummingbirds build a style of nest that differs from typical hummers; their nests are conical, frequently attached to the underside of a broad, drooping leaf.

The time required for a hummingbird egg to hatch varies considerably by location and species, from a low of about 14 days to more than a month. For species in the United States, 16 to 18 days is typical. In their nestling days, baby hummers eat a steady diet of both nectar and, significantly, insects, from which they derive critical protein to grow. The mother hummingbird usually has two chicks and must feed them frequently, especially during the first days after hatching. She gathers nectar and insects in her crop—a muscular pouch in the esophagus—and when she returns to the nest, the chicks open their mouths, heads pointing upward, and the mother injects the food mixture directly into their crops by regurgitation.

Nearly three weeks after hatching, the young hummers are ready to perch on the edge of their nest and test out their wings and very soon after will take their inaugural flights. Once they take to the wing and leave the nest, they are independent, and lucky are the hummingbird enthusiasts who, come midsummer, suddenly find their feeders and flowers abuzz with youngsters.

ABOVE Female hummingbirds build the nest and raise their young without any help from the male.

BELOW These Allen's Hummingbird fledglings are nearly ready to leave the nest.

Kids These Days

Upon fledging, juvenile hummingbirds are independent, instinctively understanding how to feed and fend for themselves. But that doesn't mean they can't learn on the job, and they do indeed learn and remember favorite feeding locations, preferred flowers, and even when they last fed at a specific bloom.

Other behaviors leave room to wonder. One year I was lucky enough to watch an unbridled juvenile male Rufous Hummer attempt to ply his charms on a female Anna's. She was perched on a twig when the bold little junior Rufous sidled up and began waggling back and forth within inches of her; she watched for a few second and then chased him away.

But our little Prince Charming was hardly finished. A day later I was situated in my photo blind, which sits near my birdfeeders, when I heard the unmistakable buzzing of a hummingbird shuttle display. I looked out the side window in time to watch the juvenile Rufous trying his luck on a Black-capped Chickadee. The Chickadee seemed unimpressed, so the little hummer darted over to a Purple Finch that was perched in the briars and again, just inches from the target of his apparent romantic notions, launched into an impressive shuttle display. The finch, like the chickadee, was insouciant. A few days later he reprised his act, to no effect whatsoever, on a Red-breasted Nuthatch. Finally, in a case of turnabout is fair play, a week later I watched a young male Anna's Hummingbird perform the species' amazing display dive several times for a young female Rufous, which watched from her perch on a wire fence.

Surprisingly perhaps, such behaviors are not limited to juvenile hummers. Males of other species have been observed performing what we interpret as display flights—dives and shuttles—aimed at other species, especially various songbirds.

Dive Bombers

Some hummingbird males, including those of several common species in North America, perform amazing courtship dive flights. They begin by gaining altitude and then plunge toward earth, accelerating to dazzling speeds before suddenly

The male Allen's Hummingbird uses spectacular dive displays to woo females.

checking their dives in front of the female they are trying to impress. Many species also, or instead, perform what are called shuttle flights or waggle flights, wherein they show off in front of the female by flying left and right in species-specific patterns.

An Allen's Hummingbird takes a quick bath at the edge of a water feature designed for birds.

In the far western United States, the males of most common species engage in this kind of courtship behavior. The Anna's Hummingbird is among them, and he not only uses high-speed diving maneuvers but also belts out his characteristic scratchy song from a favorite perch. Interestingly, Anna's Hummingbirds will sing—and nest—at any time of year given opportune conditions.

Dive displays are audible. The male's tail feathers are structured so that at a particular part of his aerial maneuvers—usually when he checks his speed at the bottom of the dive—the rush of air meeting feather produces a high-pitched *whirr*. Each species sounds different, and keen observers can learn to identify the species by the sound made during a male's dive.

Hit the Showers

Hummingbirds meticulously preen their feathers—in that respect they are like racecar drivers who take great care to keep every part of their automobile in perfect working order. Hummers love to take a shower or even a bath, using water to help clean their plumage, and as such are attracted to gardens with water features, intentional or not. One female Anna's Hummingbird in my backyard learned to use my incessantly dripping faucet for her showers by perching on the two turns of exposed metal hose threads. A few years later, a male Rufous Hummer learned to accompany me in the garden on my watering rounds with the hose; the bird would fly in and out of the spray and sometimes perch on the plants I was watering to catch the edges of the shower. Of course, hummers visit birdbaths and other such standing-water features. And a little rain doesn't stop them: hummingbirds continue to fly in the rain, shaking their already-heavy heads violently, 132 times per second, to dispel the drops of water as they go.

All Heart—and Bird Brain

Hummingbirds have the largest hearts relative to body size in the entire animal kingdom; their hearts account for as much as 2.5 percent of their body weight. To put that figure in perspective, the human heart accounts for only about 0.3 percent of total body mass. And the hummingbird's heart beats at astounding rates, typically 500 to 1,200 beats per minute, depending on level of activity. What's more, hummingbirds have the largest brains relative to their body size of any bird. Those robust brains account for 4.2 percent of their total body weight—a human brain accounts for only 2 percent of body weight. A special part of the brain involved in motion detection, the lentiformis mesencephali, is much enlarged in hummingbirds compared to other birds, including other birds that can hover for short periods of time. Researchers believe this enlarged feature aids hummers in their unique ability to maintain position in space while hovering and during high-speed flight because it allows them to detect and process movement equally from all directions, unlike all other animals.

Also, thanks to their large hippocampus, hummingbirds are expert at remembering where their favorite food sources are located, even to recalling which individual blooms they have fed from recently. This episodic recall (the ability to remember and act on past events) allows them to avoid wasting time (and energy) revisiting blossoms that they have already depleted of nectar. Episodic recall was long considered an ability unique to humans, but over the decades, researchers have realized that some animals possess it to varying degrees, hummingbirds most impressively.

The Curious Case of the Anna's Hummingbird

The Anna's Hummingbird is the most abundant hummingbird west of the Sierra Nevada and Cascades in California, Oregon, and Washington. Until the 1930s, this species' breeding range was confined to central and southern California. But then came a curious northern range expansion: the first Anna's Hummer showed up in Oregon 1944; by the 1970s, they had spread all the way to British Columbia. Those initial pioneers established a founding population, and within a few decades, Anna's Hummingbirds were ubiquitous in the Pacific Northwest. According to the International

Anna's Hummingbird routinely spends the winter in the northern parts of its range, able to withstand subfreezing weather, especially with a little help from people who maintain hummingbird feeders.

Union for Conservation of Nature, their population has doubled since 1980.

While this northward range expansion was capturing the attention of birders and ornithologists, a variety of reasons for it were posited. One theory was that a warming climate was allowing the birds to expand their range ever northward, but a research project undertaken by the Cornell Lab of Ornithology and Cal State Los Angeles demonstrated that, in fact, the northward march of the Anna's Hummingbird is all about people. Anna's Hummers had, essentially, embraced humanity: as the human population in the Northwest (and lately in the Southwest) expanded, the hummers followed along, because where people take root, so do the flowering plants we like to establish in our yards, parks, and urban centers. The Anna's Hummingbirds were simply following the gravy train, which included increasing numbers of hummingbird feeders.

An equally perplexing question arose alongside this northward range expansion: how can these tiny hummingbirds survive the cold winters of the Pacific Northwest? Granted, the concept of a cold winter in temperate western Oregon or western Washington would be laughable to a resident of the Rockies or Great Plains, but nonetheless, every Northwesterner who feeds Anna's Hummers during the winter has marveled at their ability to withstand subfreezing nights, ice, and snow.

At least partially nonmigratory in their historic range in the chaparral belt of California, Anna's Hummers expanded their range northward so rapidly that they likely never had time to evolve a seasonal escape strategy—i.e., a southward migration. They are now the northernmost year-round resident hummingbird. Because of their year-round status in the Northwest, Anna's Hummers require special tactics on the part of their human hosts who provide feeders (discussed in the next chapter), and like all northern hummers, they rely heavily on torpor to make it through the cold nights.

Watch Your Blind Side

Hummingbirds, despite their amazing speed and agility, are not immune from predation. Corvids—jays, magpies, crows, and others—are notorious for raiding songbird nests to eat eggs and chicks, and hummingbirds are given no quarter by these highly intelligent winged assailants. Hummingbird nests are also vulnerable to arboreal snakes and other creatures.

Members of the corvid family, such as this Mexican Jay, routinely pillage hummingbird nests to eat eggs and young.

But even adult hummingbirds can find their way into the maw of a hungry predator. As unlikely as it may seem, common house cats and their feral brethren are perhaps the single most widespread threat to hummingbirds. Felines are deadly and dedicated predators, capable of surprise attacks on all manner of birds, regardless of size and speed, with estimates suggesting they kill between 1.4 and 3.7 million birds annually in the United States. In fact, domestic cats, an invasive species (especially when feral), are one of the causes behind the continuing population decline of the critically endangered Juan Fernández Firecrown, a hummingbird found only on a single island off Chile.

Beyond cats, hummingbirds face threats from predatory birds such as small hawks, along with spiders, and even praying mantises. Aggressively predatory large insects, mantises will stake out flowers frequented by hummingbirds and even cling to hummingbird feeders waiting for an opportunity to make a deadly-fast strike on an unlucky hummer. Any mantis found on a feeder should be relocated a long distance away.

Praying mantises pose a threat to hummingbirds.

How to Hire a Bodyguard

A detailed and intriguing study conducted in Arizona's Chiricahua Mountains—a hummingbird stronghold—revealed that hummingbirds that nest close to active Cooper's and Northern Goshawk nests enjoyed considerably better reproductive success than those nesting far from hawk nests. These accipiters specialize in preying on other birds; hummingbirds are simply too small and agile to attract their attention, but corvids (jays, magpies, and the like) are on the menu. So, not surprisingly, the Mexican Jays that inhabit these mountains—and which happily seek and eat hummingbird eggs and fledglings—try to keep their distance from the hawks.

A Cooper's Hawk, standing guard. Hummingbirds that nest near Cooper's Hawk nests have much better nesting success than those outside the zone of protection provided by these hawks.

For the jays, avoidance is the best strategy for safety, so they not only keep their distance from active hawk nesting sites, but when near those places they also tend to fly higher. Cooper's Hawks and Northern Goshawks hunt by plunge-diving and by high-speed horizontal pursuit of their prey, often from a hidden perch but also from aloft, so the jays try to avoid flight and foraging activities that put them at risk. Fewer jays means fewer hummingbird predators and improved nesting success. The research team, which published their results in *Science Advances* (4 September 2015), monitored 342 hummingbird nests over several years. Only 20 percent of the hummingbird nests built in plots without active hawk nests achieved brood survival. In other words, the hawks serve as an unwitting security force for the little hummers: when jays alter their foraging behavior, thanks to the presence of hawks, an "enemy-free space" for nesting hummingbirds is the happy result.

ABOVE Sugar water from artificial feeders helps give hummingbirds the energy they need to voraciously hunt meat.

BELOW This remarkable image perfectly illustrates how hummingbirds "hawk" insects from the air.

Bug Out

Nectar is the high-octane nosh that fuels hummingbirds. But hummers also need body-building nourishment. Proteins, fats, and amino acids are critical to a hummingbird's survival and well-being; and they spend considerable time hunting and eating the small insects, spiders, and other arthropods that provide these vital compounds. Thanks to their amazing agility and a special adaptation that essentially makes their bills a spring-loaded set of chopsticks, hummingbirds can snatch insects out of the air; hovering, they also glean earthbound prey from spider webs, vegetation, and other places. Such insect and arthropod prey is not only crucial for adult hummingbirds but also a critical part of the diet mother hummers feed their growing chicks. I once read that hummingbirds are "sugar-powered flycatchers" (or words to that effect)—an apt description of these mighty meat-eating mites.

Sap on Tap

Flower nectar and bugs (and sugar water in populated areas) are hummingbird staples. But some hummingbirds also take advantage of another food—tree sap. This sweet oozing liquid has a high sugar content, not unlike nectar, so it's ideal for hummers—if only they could access it. Here's where sapsuckers, a type of woodpecker, come in. Sapsuckers drill holes into trees for the sap they eat, creating rows of sap wells, and some hummingbird species, including Ruby-throated, Rufous, Calliope, and Broad-tailed, feed readily on the sap in these wells. Both sapsuckers and hummingbirds will defend sap wells against interlopers.

Sapsuckers, such as this Red-breasted Sapsucker, provide access to a rich food source, one that is particularly important to hummers whose spring migration takes them through and to places where flowers are not yet blooming in abundance.

North America has four species of sapsuckers: the Far West is home to the beautiful Red-breasted Sapsucker; the Interior West has the closely related Red-naped Sapsucker along with the Williamson's Sapsucker. The Yellow-bellied Sapsucker reigns throughout the East, and breeds from New England and the Canadian Maritimes across the northern Plains, all the way to the east slope of the Rockies in Alberta and British Columbia, and northward to the Yukon and Northwest Territories. Its winter range is equally massive, from the Mid-Atlantic to eastern Colorado and southward throughout the East and throughout Mexico and the Caribbean.

Significantly perhaps, the breeding range of the Ruby-throated Hummingbird falls entirely within the range of the Yellow-bellied Sapsucker. In the West, sapsuckers occur in all forested montane areas, and, especially in the case of the Red-breasted Sapsucker, many lowland regions. They require forests, and while the Red-breasted often lives in deciduous stands, the Red-naped and Williamson's Sapsuckers are strongly tied to coniferous zones and mixed stands. These same places are home to the hummers of the northern mountains—Rufous, Broad-tailed, Black-chinned, and Calliope. Western sapsuckers are largely migratory: Red-breasted Sapsuckers overwinter in the Northwest but withdraw from their northernmost breeding range in British Columbia; Red-naped Sapsuckers spend the winter from the Desert Southwest to central Mexico; other than some year-round range in California,

Williamson's Sapsuckers likewise head south. However, sapsuckers return north earlier than hummingbirds, and in this way, help pave a food trail for northward-bound hummers.

What's in a Name?

Throughout the avian world, some birds have simple descriptive monikers, others more curious names. The pragmatic naming paradigm gave us the Ruby-throated, Violet-crowned, and Black-chinned Hummingbirds, for example. But who were Anna and Allen, Costa and Rivoli? And why Lucifer and Calliope? All these names date to the early to mid-1800s, when the study of New World birds, and especially hummingbirds, progressed at a quick pace.

The Anna's Hummingbird honors Anna Masséna (1802–1887), the Princesse d'Essling and Duchess of Rivoli, in Paris. Anna's husband, Francois Victor Masséna (1799–1863), was an amateur ornithologist with a substantial collection of preserved hummingbirds, including a little beauty whose crown and throat were adorned in shiny magenta tones. But it was another Parisian, René Primevère Lesson, who, having studied specimens of this bird in the duke's collection, named it after Anna.

René Primevère Lesson studied and named numerous hummingbird species; in 1832 he presented beautiful artistic renditions of them in his *Histoire Naturelle des Colibris* (Natural History of Hummingbirds).

Lesson also described Allen's Hummingbird (as *Ornismya sasin*) in 1829. Some 50 years later, ornithologist Henry Henshaw, unaware that the species had been named and then described in the 1837 edition of *Dictionnaire Classique des Sciences Naturelles* by Belgian naturalist Pierre Auguste Joseph Drapiez, studied specimens collected by Californian Charles Allen. At the time, the look-alike Rufous Hummingbird was well known to science, but although an amateur, Allen was a keen observer. The hummingbird skins he originally sent to ornithologist William Brewster in Massachusetts included his notes about the differences in the tail feathers between the green-backed specimens and the rufous-backed specimens; Allen suggested the two might be different species, and Henshaw agreed with him, naming the new species *Selasphorus alleni*. Ultimately something of a compromise led to re-adoption of *sasin* as the specific epithet, with Allen's Hummingbird as the common name.

The flamboyant Costa's Hummingbird is named for Louis Marie Pantaleon Costa, Marquis de Beau-Regard, an astute amateur naturalist who began collecting birds at age 15. He was especially fond of hummingbirds and eventually compiled an extensive collection of specimens. His friend Jules Bourcier named the bird, having studied a specimen or specimens collected by Adolphe-Simon Neboux in 1837, when he served as ship's surgeon aboard the *Vénus*; at the time of Neboux's discovery, the French frigate was anchored in Bahia Magdalena, Baja. Costa's Hummingbird, like the

The beautiful little Costa's Hummingbird of the Desert Southwest was named in honor of French statesman Louis Marie Pantaleon Costa (1806–1864), an avid hummingbird collector.

Anna's, is a member of *Calypte*, the capped hummingbirds. Fittingly the genus name derives from the Greek *kaluptre* ("veil," "veiling"), aptly describing the crown and gorget of iridescent feathers on both species.

Rivoli's Hummingbird is another species that owes its moniker to Lesson. At the time (1829), Lesson was not aware that two years earlier, the British ornithologist and artist William John Swainson (1789–1855) had described this same bird. Although Swainson had collected some 760 birds skins on a multiyear expedition to Brazil in the early 1800s, he may have first studied this species from specimens already in Great Britain. At least one writer suggests Swainson also traveled to Mexico, which unlike Brazil, is home range to Rivoli's Hummers. In any event, Lesson, according to William Jardine's *The Natural History of Humming-birds* (1833), used the more elegant name, Duke of Rivoli's Hummingbird (that is, Anna's husband, the aforementioned Francois Victor Masséna).

Lesson tried again to attach Swainson's name to another hummer first described by that Englishman, the beautiful Lucifer Hummingbird, calling it *Trochilus swainsonii*, but Swainson had already applied the name *Calothorax lucifer* to this species. The name Lucifer (Latin, "light-bearing") recalls the fallen angel of Judeo-Christian tradition, but Swainson no doubt had the male's brilliant gorget in mind when he bestowed the moniker.

Finally, the dainty Calliope Hummingbird is named for Calliope, the Greek muse of poetry. This mountain specialist of the Far West was first described by English ornithologist and bird artist John Gould (1804–1881) in 1847; two Americans had sent him specimens, helping him to realize that this species "ranges very far north [from Mexico]," as he notes in his *Introduction to the Trochilidae, or Family of Humming-birds* (1861). Many decades would pass before, late in the 20th century, ornithologists decided the Calliope Hummer belonged in the genus *Selasphorus* ("light-bearing"), alongside the Rufous, Allen's, and Broad-tailed Hummingbirds—certainly a less alluring moniker than its original genus, *Stellula* ("abounding in small stars").

ABOVE John Gould published a series of monographs, with beautiful artwork by his wife, Elizabeth, and other artists; the volumes are masterpieces and extremely valuable. Gould displayed his collection of 320 hummingbird species at the Great Exhibition of 1851 in London but did not see his first hummingbird in the wild until traveling to the United States in 1857.

BELOW The Calliope Hummingbird is one of several species found in the Northern Rockies.

In Central and South America—the hummingbird heartland—many dozens of species bear evocative names, sometimes descriptive and sometimes rhapsodic, bestowed by the spellbound observers who first studied these entrancing birds. Hence we have the Woodstars, Sunangels, Mountain-gems, and Brilliants, mixed in with Trainbearers, Pufflegs, Racket-tails, and Plovercrests. The variety is joyfully bewildering.

Natural habit and habitat cycles can explain why hummingbirds sometimes suddenly disappear from your yard.

Disappearing Act

In northern latitudes, migratory hummingbirds suddenly appear one spring day to the delight of people who eagerly anticipate their annual arrival with feeders full and waiting. Along key migration routes and in nesting territories, hummingbird activity can be frenetic during those first few weeks after the birds arrive. Some birds may be using your feeders to fuel up for the continued northward journey; others establish and defend feeding and breeding territories in your yard and around your neighborhood. But in some places, seemingly overnight, the birds disappear or considerably thin out, often puzzling their human hosts.

In many cases, this is because migrants have moved on; in other cases, once a male begins defending its territory, he may drive away rivals. During the breeding season, your local females may choose nesting sites away from your yard and feeders—they seek nest locations that are well concealed and hopefully safe from predation; once females are nesting, they spend a great deal of time sitting on eggs, leaving only to forage for food as needed. When the eggs hatch, the female must feed the chicks insects alongside nectar, so she may well be a less frequent visitor to feeders. Then come midsummer, males typically begin departing the breeding territories—substantially earlier than the late-summer southward migration of females and their young.

Moreover, natural habitat cycles can affect local hummingbird densities. In some places, years with exceptional wildflower blooms may disperse breeding hummers over a broader area. Also, as the season progresses, food sources change—new species of flowers bloom as other fade, and new species of insects emerge and become available to foraging hummers. These changes lead the birds to new locations within or near their territories, or even draw postbreeding birds to new areas. Likewise, the availability and quality of hummingbird feeders plays a role in where hummingbirds are found; as discussed in the next chapter, hummingbird feeding is a competitive endeavor: you must be better at it than your neighbors to assure a steady supply of hummers at your feeders.

So, if your hummers seem to vanish after their exuberant spring arrival, often you can chalk it up to natural cycles. Hedge your bets by being a great hummingbird host. Then hope for renewed activity when the youngsters are out of the nests—and the southbound migrants stop to refuel.

Migratory hummingbirds such as this Rufous must fuel up for the long journey to their wintering grounds in Mexico, so leave your feeders up until well after you've seen the last hummer of fall.

Vacation in Mexico

Hummingbirds that live in northern latitudes routinely withdraw for the winter, migrating southward, mostly to Mexico. Such migration is the typical life history for Ruby-throated, Rufous, Black-chinned, Broad-tailed, Calliope, Costa's, Allen's, and Anna's Hummingbirds, the eight common species in the United States. Anna's Hummers also occur as year-round residents in a significant part of their range, and Allen's Hummers are year-round residents in southernmost coastal California. The other species found in the United States are largely confined to southern Arizona, southern New Mexico, and southwestern Texas at the northern extent of their summer range, but they too tend to withdraw southward into Mexico for winter.

For decades, persistent folklore suggested that hummingbird feeders needed to be taken down in early fall. That way, hummers would not be tempted to stick around in northern latitudes as the approaching winter ushered in ever-colder weather that is decidedly unfriendly to the little dynamos. But you need not worry: the genetic drive to migrate is far more powerful than the ephemeral draw of a favorite feeder. So leave those feeders up—the soon-to-be migrants need to fatten up for the long flight ahead. Only when they have gone should you remove the feeders for the winter—unless you live on the West Coast, from California to British Columbia, where the Anna's Hummer is a common winter resident.

Prepare for the spring return of the hummingbirds by having their feeders ready and waiting. The timing of the arrival depends on location and species. Naturally, northbound migrants arrive in the southern portions of their range first, and some species migrate earlier than others. Some great online resources can help you estimate when your vacationing hummers will be back. One of the best is eBird (ebird.org) by the Cornell Lab of Ornithology: you can select the "Explore" tab, choose "Species Maps," and then fill in the species, a date range, and a location. For example, if I want to know when to expect Broad-tailed Hummingbirds in Aspen, Colorado, I select that species and location. Then, knowing that they are spring arrivals, I can choose April as the timeframe and quickly see that very few sightings have been reported in April. But when I select May, sightings are ubiquitous, and each eBird drop-pin on the map shows the date of that sighting. Looking at the dates for a dozen or so May sightings, it quickly becomes apparent that if I lived in Aspen, I'd want my hummingbird feeders ready and waiting for the return of the vacationers in the first week of May.

To Kill a Hummingbird

With all due respect to Harper Lee, to kill a mockingbird would be far easier than to effect the demise of a hummingbird. These incredibly swift, amazingly agile birds are physically adapted for a high-speed lifestyle, and as adults, they are well able to evade would-be predators. Or at least most of them: when James Cook set anchor in Nootka Sound, in what is now British Columbia, in 1778, he had with him naturalist Joseph Banks, who collected many bird specimens. Among them were Rufous Hummingbirds; Banks reported that "natives brought them to the ships in great numbers."

How the indigenous people could manage to supply copious quantities of hummingbirds to Banks (no doubt in trade) is at least partially subject to conjecture, but they may have used the same method they used to capture the poor little birds to serve as toys for their children. In *The Northern and Central Nootkan Tribes* (1951), Philip Drucker reports that these people trapped hummingbirds "by smearing slug slime on twigs in the vicinity of flowering plants. The unfortunate birds so captured were fastened to a string, threaded through their nostrils, so the children could play

Native Americans had ingenious methods for capturing hummingbirds, and, as recorded in Codex Mendoza in the early 1500s, Aztec elite made prolific use of bright feathers in their ceremonial dress.

with them, making them fly round and round." Likewise, Grant Keddie, curator of archeology at the Royal B.C. Museum, has seen references to children catching hummingbirds on the ends of sticks covered with tree sap. Apparently, they had a vested interest: hummingbird skins were often used to decorate children's clothing in the province's interior. We cannot know whether sticky substances on sticks was a universal method of catching hummingbirds throughout the Americas, but where hummers abound around favorite flowers—such as wild currant bushes in British Columbia—such a method would have been highly successful.

The only other way I can imagine indigenous people catching hummingbirds would be with nets. To do so would have required large nets of extremely fine fiber and small mesh size, or easily wielded long-handled bag nets. Native American fiber and net technologies were highly advanced but—compared to stone artifacts—tremendously underrepresented in the archeological record: stone survives for millennia and beyond, but organic materials usually rot away in no time unless luckily preserved in substrate. Thus, while the term Stone Age is nice shorthand, a more apt moniker for the pre-contact period of indigenous cultures would be something like Stone/Bone/Wood/Plant/Sinew Age. Native cultures, which have occupied the Americas for at least 14,000 years, developed incredible fiber technologies, and I have little doubt they possessed the skill and raw materials to make netting fine enough to capture hummingbirds.

The original inhabitants of coastal British Columbia might have had no misgivings about killing hummingbirds, but this was not the case throughout the Americas. One Mayan legend holds that not only is capturing and caging a hummingbird essentially impossible, but to do so would lead inevitably to bad luck for the perpetrator. These days, killing or capturing hummingbirds is expressly forbidden by federal law in the United States and most other nations. They are protected domestically by the Migratory Bird Treaty Act, as are jays—and mockingbirds.

How Far Can You Fly?

In January 2010, licensed hummingbird bander Fred Dietrich captured and gently banded a female Rufous in Tallahassee, Florida. Five months later, Kate McLaughlin, another hummingbird bander, recaptured the little hummer—across the continent at Chenega Bay, Alaska. Establishing a new long-distance migration record for hummingbirds, this little Rufous may have flown some 4,000 miles between the two points: 3,530 miles is the straight-line distance, but Rufous Hummers don't migrate in a straight line.

So, a Rufous Hummingbird may be the long-distance champion, but all northern-latitude hummers migrate to one extent or another. The northernmost species—Rufous, Calliope, Black-chinned, Broad-tailed, Allen's, and Ruby-throated—all undergo substantial twice-yearly journeys to and from their breeding grounds. Many Ruby-throated Hummers annually gather to fuel up on the Gulf Coast and then depart for what must be a perilous long-distance journey over open water: they fly across the Gulf of Mexico to their winter range in Mexico and Central America. And then they repeat that journey in the spring.

Speaking of Rufous Hummingbirds, these little dynamos are the northernmost hummer species, breeding as far north as the Gulf of Alaska and Prince William Sound. Only the range-expanding Anna's Hummer and the occasional vagrant Costa's Hummingbird approach such northern latitudes, and then only as rare casual visitors. But what is the southernmost species of hummingbird? That distinction belongs to the Green-backed Firecrown of Chile and Argentina: this little hummer ranges as far south as Tierra del Fuego, at the very tip of South America. As the crow flies, some 9,000 miles separate the northernmost range of the Rufous

The diminutive Green-backed Firecrown is the southernmost hummingbird species.

Hummer from the southernmost range of its extreme-conditions soulmate.

On the Wings of Geese

Another persistent myth: hummingbirds, too tiny and fragile to make 1,000-plus-mile migrations on their own each spring and fall, hitch a ride on migrating geese or other large birds. The source of this old tale is elusive, but in addition to being demonstrably false, it is simply silly. One version has the tiny creatures burrowing into the warm, soft feathers of the host for the long ride; another versions suggests they ride on the goose's back, holding on (for dear life, apparently) with their slender bills. I've heard yet a third variation, which holds that the hummers wedge themselves tightly up against the goose's body, where its powerful wings join its torso. No doubt other permutations of the myth exist as well, but a myth it is.

Ever so briefly, however, I had to reconsider the potential veracity of this old fable. One mid-March afternoon, a flock of some three dozen Canada Geese flew overhead while I was outdoors, and amid them was a hummingbird, buzzing acrobatically from goose to goose as if trying to drive them away—or regain his purchase snuggled within a goose's downy coat. Of course I didn't really believe the latter, and I don't know why a hummingbird would try to intimidate geese on the wing, but witnessing such interaction certainly underscored for me the many behavioral traits that make these tiny bundles of energy so fascinating.

Hummingbirds swarm around a red fuchsia. Red flowers are especially attractive to hummingbirds, but ultimately, nectar content and quality matters more than color.

Seeing Red

Hummingbirds have superb visual acuity; they see color even better than we do, their vision extending into the ultraviolet spectrum. But their eyes are adapted to see warm shades better than cooler shades. This ability to easily pick out red, orange, and yellow flowers amid a sea of cool green led to the long-held assumption that they prefer red to all other colors. Scientists have put that assertion to the test in various ways, however, and have learned that the richness of the nectar matters more than the color of its source. In other words, hummingbirds are quickly able to switch from a preference for flowers in the red to yellow range to a preference for the most nectar-rich flowers, regardless of color. They are quick learners, and it is nourishment they are after.

In an early experiment of this sort, published in 1980, Michael Wheeler used his local Anna's Hummingbirds as test subjects. He found that the birds keenly recalled exact locations of food sources, and when Wheeler moved the feeders, the hummers swiftly relocated them. He also placed five feeders side by side, each with a different color of nectar but otherwise identical, and he shifted their order randomly to avoid location bias. He found that once a nectar source was located, the color of the source had no effect on the number of birds approaching it, or feeding from it; moreover, no matter the color of the liquid solution, the same number of drinks were taken per unit of time. However, his subject hummers approached, and fed from, red containers more frequently than yellow, green, clear, and blue containers, in that order. Ultimately, after quantifying all his data, Wheeler determined that, at least for these Anna's Hummers, feeding behavior was most influenced by time of day, weather, and familiarity with the location of the food source, respectively, and only then by the color of the source.

Such findings offer two takeaways for the backyard enthusiast who wants to attract hummingbirds. First, create a colorful environment to draw them in by using colorful flowers and colorful feeders—that most commercially available hummingbird feeders are red or feature red parts is no coincidence. And second, once you have hummers using your feeders and flowers, keep them there by maintaining those feeders and flowers.

One thing you don't need to do is add red coloring to your feeder nectar, and in fact one sure way to make hummingbird experts see red is to use anything other than a simple mix of sugar and water. Based on studies conducted on other organisms, some red dyes have been correlated with decidedly unhealthy effects, and while no such studies have specifically inquired into the potential deleterious effects of red dyes on hummingbirds, the inference is sufficient: do not add red dyes to hummingbird food, and do not purchase any commercially available hummingbird food that contains dyes or other chemicals. Doing so is unnecessary: hummingbirds are clever about finding feeders meant for them, and all you need and want in those feeders is sugar water. The red-colored parts of a hummingbird feeder are sufficient to draw in the birds.

Alright Already!

Just about everyone who routinely maintains hummingbird feeders and hummingbird habitats enjoys the occasional close brush with the tiny buzz balls. Hummingbirds can seem almost inquisitive, hovering nearby as if to inspect what you are doing. They often behave this way as you refill their favorite feeder, barely able, it seems, to contain their impatience. They will hover inches from your

Close encounters with hummingbirds are common. The little speedsters routinely buzz humans and often hover very close, as if out of curiosity.

face, zip past you at warp speed; once, while pouring sugar water from a jug in my left hand into a feeder held in my right hand, a female Rufous flew between my two arms, barely missing my face. I finished the refill with all due haste, and she was back to lapping up the nectar as soon as I took three steps back from her feeder.

Of course, thinking of hummers as inquisitive or impatient risks anthropomorphizing them, when in fact, in these kinds of encounters, the tiny birds are simply behaving naturally: they habitually check out potential food sources, and they routinely defend their preferred food sources, whether flower or feeder. Such inquisitiveness and aggressiveness is preprogrammed: they can't help their little selves. These behaviors also vary considerably by species. In my neighborhood, the local Rufous Hummers act in the most belligerent way, frequently buzzing humans near the feeders, while the Anna's are more prone to hover at discreet distance and wait for a feeder to be refilled.

ABOVE The Green Hermit is one of about three dozen species of hermits.

BELOW The Tawny-bellied Hermit inhabits the mountains of Colombia, Ecuador, and Peru.

Of Hermits and Hummers

Hermits are a subfamily (Phaethornithinae) of the hummingbird family (Trochilidae) comprising about three dozen species, all found in the tropics and subtropics. They differ from other hummingbirds in that, as their name suggests, they largely lack showy iridescent colors other than green, tending instead to be inconspicuous shades of tan, brown, rust, gray, and bronze. Hermits are also less sexually dimorphic than most hummingbirds: male and female of each hermit species are similarly colored. Many hermits bear a certain subfamily resemblance to one another, with distinctive dark eye stripes bordered by contrasting light-colored eyebrow stripes. To varying extents, depending on species, most hermits form leks during the breeding season: a lek is a communal display site occupied by several males, all displaying for the attention of females.

Hermits have interesting bills—the Saw-billed Hermit and the Hook-billed Hermit, for example, but in these two cases, the reason for the remarkable bill structure has proven elusive. Most hermits have long, decurved, somewhat robust bills compared to typical hummers, and two species—the White-tipped Sicklebill and the Buff-tailed Sicklebill—have deeply curved bills specially adapted to gather nectar from heliconias, centropogons, and other beautiful tropical flowers.

Playing Hard to Get

Speaking of hermits and heliconias: researchers Adam Hadley, Matt Betts, and W. John Kress, while studying the effects of deforestation on plants and the hummingbirds that pollinate them, discovered that *Heliconia tortuosa* in Costa Rica had evolved a strategy to prefer pollination by a very specific pair of hummingbirds.

Researchers discovered a remarkable relationship between two species of long-billed hummingbirds and a beautiful heliconia in Costa Rica.

The team had struggled to successfully hand pollinate this heliconia in the lab, and then, through controlled experiments, discovered that the plant held its nectar prize deep in the flower tube—easy for only the Green Hermit and the Violet Sabrewing, both long-billed hummingbirds, to access. The researchers surmised that the plant benefits from this adaptation because these two species of hummers carry pollen over longer distances in their foraging than other hummers, reducing the chance of inbreeding from pollinating nearby and near-kin flowers and thereby improving the chances for healthier heliconia offspring.

Bug or Bird?

No other avian acrobat can match the hummingbird's incredible feats of flight, but these enigmatic little creatures do have counterparts in the insect world. Exemplifying the process of convergent evolution, many moths in the family Sphingidae feed by hovering at flowers. Moreover, these Hummingbird Moths (aka Sphinx Moths, Hawk Moths) are quite large, a robust 2 inches long or more, depending on species, and with wingspans of 2 to 5 inches. Casual observers easily confuse hummers with their moth counterparts, which feed by unfurling a long tonguelike proboscis to reach into flower tubes to gather nectar. Behaviorally, hummingbirds and Hawk Moths are so similar, that, according to a 19th-century newspaper account, "The natives in some parts of South America have a theory, indeed, that the moth is only, as it were, the larva of the bird, and that as it grows older it gradually acquires bones and a beak."

The robust Hummingbird Moth is one of the few types of animals that can hover feed, a behavior it shares with hummingbirds.

Band of Banders

Data collected by bird banders has revealed critical insights into the lives of birds. Using various means of capture, they place harmless metal leg bands on birds of all kinds, not just hummingbirds. Each band has a code stamped into it, so that when a bird is recaptured (or harvested by legal hunting in the case of game birds), scientists can gain details about the bird's movements, life span, and health. Banders catch most birds with nets, especially what are called mist nets, made from a very fine mesh that the birds can't see. Birds that fly into the mesh are gently removed, banded, and released.

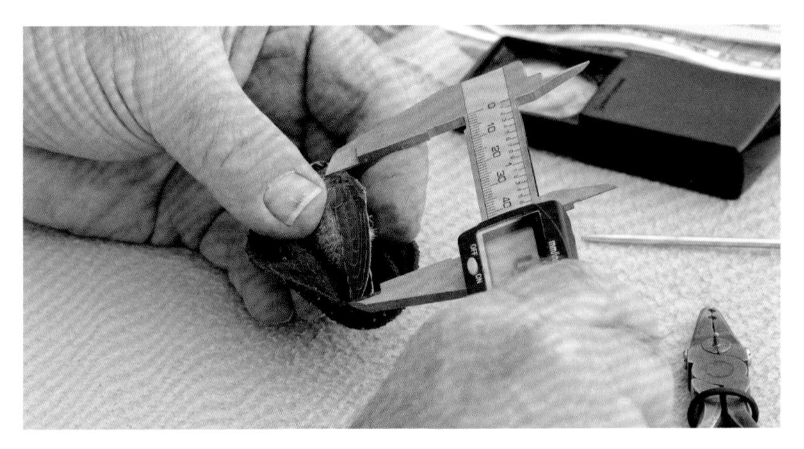

ABOVE Trained hummingbird banders carefully examine captured hummers to record details about their health and age.

BELOW A minuscule band attached to a hummingbird's leg is inscribed with a number, so that the bird can be identified at its next encounter with humans.

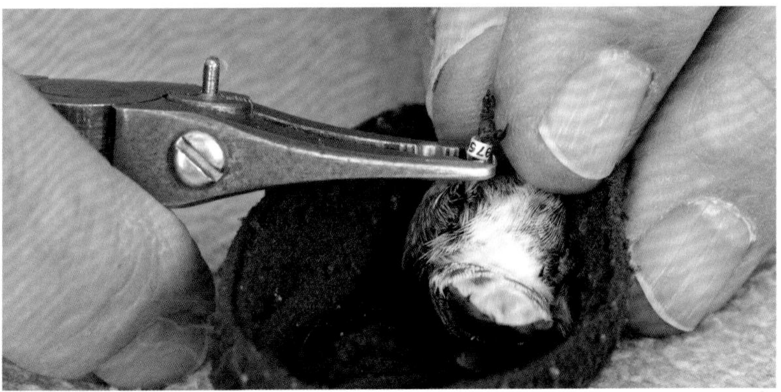

Banding tiny hummingbirds requires specialized knowledge and techniques. Banders, specially trained and then licensed by the federal government, trap hummingbirds in drop nets or fine-mesh cages placed around hummingbird feeders. Captured birds are transferred to soft mesh bags and then quickly examined, banded, and released. Banders record physical data of each bird, determine its species and age (not always easy with juveniles and females), and carefully attach the tiniest of metal identification bands on its leg. Each band has a different code composed of letters and numbers, and all information about a banded hummingbird is cataloged alongside the band number. Any time a banded hummer is recaptured, banders and ornithologists can then compare data about that bird with data from all previous encounters. In short, every banded hummer is a traveling research project, carrying with it the potential to increase our understanding of hummingbird behavior, populations, and natural history.

Larger hummingbirds, like our long-billed friend the Violet Sabrewing, likely live as long as any other hummingbird, but scientists have little data on the longevity of most species.

One Life to Live

For a long time, ornithologists assumed that hummingbirds were very short-lived, a few years at most, but banding data revealed otherwise. The oldest known hummingbird was a female Broad-tailed banded in 1976 at Colorado's Rocky Mountain Biological Laboratory at about one year old and then recaptured at the same place in August 1987, when she was more than 12 years old. She was in excellent condition upon her recapture, so presumably she could have lived a while longer. Because larger animals tend to have longer life spans than smaller animals, the age of this "grand old lady," as the scientists at the laboratory called her, surprised everyone.

Since then, banding studies have revealed some additional surprising life spans among different hummingbird species: a Rufous Hummingbird from British Columbia just one month shy of nine years old, a 9-plus-year-old Ruby-throated Hummer recaptured in West Virginia, a 9-plus-year-old Broad-billed Hummer from Arizona, an 11-plus-year-old Black-chinned Hummer, an 11-plus-year-old Buff-bellied Hummer recaptured in Texas, and an 11-plus-year-old Rivoli's Hummer from Arizona. Eight-plus-year-old specimens of Anna's, Calliope, and Costa's Hummingbirds have also been documented. So, while the average life span of a hummingbird may be just a few years, the maximum life span is a great deal longer than scientists once believed.

It's Not Over 'til It's Over

You would think that by now, with all this careful banding and monitoring, hummingbirds would have nothing to fear from humans. Sadly this is not the case. Habitat loss, especially through deforestation, threatens many species. Moreover, in Latin America, some people believe hummingbirds are imbued with magical powers, and dead hummingbird love charms support an unfortunate black-market trade. Likewise, dead-and-preserved hummingbirds are considered by some to be good luck charms for travelers.

This Ruby-throated Hummingbird died after striking a building in Washington, D.C.

By 2010, U.S. Fish and Wildlife Service (USFWS) enforcement agents were fully aware of the illegal importation of and trade in hummingbird charms. Most birds in the United States are fully protected by the Migratory Bird Act, and international trade in many species, both domestic and foreign, is prohibited or strictly regulated by the Convention on International Trade in Endangered Species; a host of other federal and state laws are designed to curb the illicit killing of and trade in protected species. But demand is a powerful motivator, Mexico is not yet on board, and hummingbird charms continue to be smuggled into and sold in the United States. Federal wildlife enforcement agents work tirelessly on this and may other issues.

Hummingbird species with very small geographical ranges or highly specific ecological niches are especially vulnerable to the impacts of habitat loss, and further pressure on fragile populations by indiscriminate harvest only hastens their potential demise. The average hummingbird enthusiast can perhaps most effectively help by lobbying elected officials to always effectively fund agencies that enforce wildlife laws, particularly the USFWS, as well as supporting organizations that actively help in the protection of wildlife and habitats.

TWO

Hummingbird Basics

They don't exactly pull up to the curb in stretch limos, but hummingbirds have reached celebrity status; they bring the bling, and they even have their own paparazzi in the ever-increasing ranks of photographers who count the radiant little birds as favorite subjects. People from all walks of life maintain hummingbird flowers and feeders simply to share in the sheer enjoyment, entertainment, and wonderment of watching these mystifying little rock stars. And it's a competitive world out there, with neighbor pitted against neighbor, trying to outdo one another in attracting hummingbirds to the yard.

The hummers have choices. If they find food and habitat more to their liking down the street, that's where they'll hang out. Luckily, there are plenty of hummers to go around, and the popularity of feeding them has even caused the birds to alter their natural behaviors. They congregate at feeders, and in gardens created for them, and they are drawn to urban habitats because of feeders and flowers—historically hummingbirds are not urbanites and not particularly social among

Ruby-throated Hummingbird.

An ideal human-made hummingbird habitat, regardless of size, includes nectar-rich flowers and at least one feeder, along with cover for roosting and nesting.

their own kind. But we hummingbird fans have changed that, and the birds themselves have proven adaptable.

Wander through any city residential area and look for hummingbird feeders, and you might be surprised at their ubiquity. They hang from front porches, from shepherd's hooks in gardens, from tree limbs, even from upper-story apartment balconies. Not long ago, in the downtown business district of my city, I spied a hummingbird feeder hanging from a tiny second-story balcony overlooking nothing more than an alley. Such popularity means you must keep pace with the competition and be a very good hummingbird host to keep the hummers coming. Game on!

Happy and Healthy: Care and Feeding

Attracting hummingbirds to your space can be as complex as redesigning your entire yard or as simple as maintaining a hummingbird feeder or two, filled with sugar water. Hummers thrive on nectar from a variety of flowers, the best of which are discussed in the next chapter. They also need habitat—places to perch, so they can survey and defend their territories; places to roost in safety; places to build nests and raise young; and places to hunt tiny insects for protein; they also need water, sometimes for drinking but also for grooming their plumage.

If you have a yard that can accommodate all these needs, and the time to cater to hummingbirds, you stand a great chance of attracting them. But don't be deterred if your space is smaller, or you don't have time to maintain a complex hummer

LEFT + ABOVE Hummingbirds need perches from which to survey and guard their territories. This male Anna's took over a wire fence adjacent to its favorite flowers while a Black-chinned Hummer chose a favorite twig, .

habitat: put out one or two hummingbird feeders, well cared for, and you may well find yourself serving as a hummingbird host much to the delight of your family and friends.

Unsurprisingly, rule number one for feeding hummingbirds is that natural nectar sources are best—that means flowers that the birds love. Natural nectar has nutrients that are lacking in sugar water. Still, hummingbird feeders provide a wonderful means of allowing close observation of these fascinating creatures, and such observation leads to appreciation which, I hope, then leads to stewardship and perhaps even a deeper desire to understand the broader ecology around us. For that reason, I encourage people to use hummingbird feeders—but only if they are kept immaculately clean, filled only with fresh sugar water, and refilled frequently with fresh sugar water.

That's rule number two: feeding sugar water to hummingbirds carries with it an ethical

Saucer-style feeders are easy to fill and easy to keep clean.

responsibility, which begins with keeping those feeders meticulously clean and filled only with the freshest sugar water. Sugar water can ferment very quickly, especially in warm weather. Fermented sugar water ingested by birds enlarges their livers, imperiling their health. Sugar water is a nursery for bacteria and for mold, potentially dangerous pathogens for hummingbirds. So above all else, change out the sugar water in your feeders frequently and clean the feeders themselves assiduously.

Hummingbird feeders come in two basic types. The first, saucer-style feeders, are dishlike reservoirs with removable lids and a central stem with a hook at the top for hanging (they are also available in suction-cup window-mount versions). Fill the reservoir with sugar water, replace the lid, and hang the feeder out for the birds. Among the best is the popular HummZinger from Aspects. A Michigan company (thehummingbirdfeeder.com) makes saucer-style feeders with multiple raised conical ports on the lids; the design eliminates bee/hornet access while offering the easy-cleaning advantage of saucer-style feeders.

The second basic type of feeder is the bottle-style feeder, a plastic or glass vessel with feeding ports on the bottom. Saucer-style feeders tend to be easiest, but bottle-style feeders often have more capacity. Many bottle-style feeders fill from the bottom (flip the reservoir upside down, twist off the section with the feeding ports, refill, reattach the feeder-port section, and return the unit to right-side up); others fill from the top with saucer-style reservoirs at the bottom. In either case, they can be messy, but high-capacity bottle-style feeders are handy if you host a lot of hummingbirds.

In warm weather, and definitely in hot weather, replace the sugar water in your feeders every day. That's a substantial commitment, but clean feeders with fresh food avoid the pitfalls of pathogens being ingested by hummingbirds, and while many

RIGHT The iconic Perky-Pet bottle-style feeder has been a favorite for decades. Launched in Colorado in 1958, the company now makes many different models and styles of feeders and is a prominent supporter of the Hummingbird Society, bird research and education, and various bird festivals.

BELOW Bottle-style feeders come in many different styles.

ABOVE Artistically rendered glass hummingbird feeders are available in many styles.

RIGHT + BELOW PAR-A-SOL hummingbird feeders set the standard for artisan-crafted glass feeders. This renowned company offers many different designs, from relatively simple globes to intricate multi-port feeders.

sources recommend changing out the sugar water every few days in hot weather, I encourage hummer enthusiasts to be more fastidious and do so daily. The best time to clean feeders and add fresh sugar water is in the evening, after sunset, when the hummingbirds have gone to roost for the night. That way, you'll have fresh food waiting for them first thing in the morning when they need a boost to kickstart their day.

Feeders placed in well-shaded locations out of direct sunlight can last two, maybe three days, in warm weather, but be vigilant and always err on the side of cleanliness: if you wouldn't drink out of your hummingbird feeder, neither should the hummingbirds. If you need to be away from home for more than two or three days, ask a family member or neighbor to empty your feeders, rinse them thoroughly in hot water, refill them, and rehang them.

Yeast infections in hummingbirds are frequently fatal because the condition causes their tongues to swell, meaning they cannot eat, and they quickly starve to death. Ten percent or more of the hummers who wind up in rehabilitation centers have yeast infections from improperly maintained feeders. In short, if you don't have time to properly maintain hummingbird feeders, don't have them at all. Hummingbird-attracting flowers are easier to maintain for busy people.

When you buy your first hummingbird feeders, also buy basic cleaning supplies for them: a set of two or three bottle brushes, including a fine-diameter brush for cleaning out feeding ports; a basic scrub brush of the type used to wash dishes; a bottle of bleach; pipe cleaners, which are great for cleaning tiny feeding ports and other small nooks and crannies in some feeders; and kitchen-style rubber gloves to protect your hands from bleach.

To thoroughly clean a hummingbird feeder, first disassemble it. That may be as easy as removing the lid from a saucer-style feeder, or as complex as removing the feeding ports and bee guards from various kinds of bottle-style feeders. In a large bowl, mix a mild bleach-and-water solution—about 10 percent bleach. Running hot water in your sink, use the bottle brushes and scrub brush to thoroughly scrub all parts of the feeder. Then immerse the parts in the bleach solution for a few minutes, remove, and rinse repeatedly in hot water. If your feeder has a screw-off assembly containing the ports, allow hot water to run through it, gushing out the ports, for several minutes.

Some people espouse the use of vinegar or mild detergent rather than bleach. Any of these methods are fine, but extremely thorough rinsing is critical regardless of what cleaning agent you use. Rinsing is the most important step: a clean feeder—with no mold *and* no bleach, vinegar, or soap residue—is what you're after. If you clean the feeders frequently—every couple of days when you refill them—scrubbing with hot water alone will deter mold growth. If you go more than a few days, or if you see any sign of black mold in the sugar water or on the feeding ports, use the bleach solution.

Another great option is to run feeders and their parts through the dishwasher, provided they are dishwasher-safe, or even hand-cleaning and then running them through the dishwasher with no detergent added as an effective final rinse. For bottle-style feeders, place the open end of the jar downward so hot steam can fill the vessel. Don't open the dishwasher until the dry cycle is complete to take full advantage of the hot internal temperatures, which sanitize the feeder.

Cleaning hummingbird feeders may sound like a lot of work, but after a few practice runs, you'll have the process down to just a few minutes per feeder. Moreover, if you keep them clean, and only filled with fresh sugar water, the feeders won't build up black mold or other debris.

Feeder placement is critical to the longevity of the sugar water. Avoid sunlight to the extent possible because sun and heat are the enemies of sugar water, causing it to spoil quickly. Hang your feeders in at least partial shade; protecting them from the afternoon sun is especially important. The exceptions to this are the Pacific Northwest, within the year-round range of the Anna's Hummingbird, and any place where hummingbirds arrive early enough in the spring that they must contend with cold weather. During the dead of winter in the Northwest, Anna's Hummers appreciate a feeder that is basking in the meager midday sun—and that sun helps prevent the sugar water from freezing.

For the sake of convenience, especially during hot weather when sugar water spoils quickly, consider rotating two feeders for each location: while one of the feeders serves the birds, the other is inside, clean, ready to be filled with fresh sugar water. At the end of a day or two, swap them out—then you have some time to clean the replaced feeder and fill it with fresh sugar water.

Now about that sugar water: always use four parts water to one part sugar. Boil the water in a pot, remove it from the heat, then stir in the sugar until it dissolves; allow to cool, then fill your feeders. You can refrigerate extra sugar-water mixture for about a week, but after that, discard it and make a fresh batch. Never put out hot sugar water. If you've run low and the birds are knocking at the door, you can cool the mixture more quickly by boiling two cups of water, then mixing in the sugar, then adding two cups of cold water, and allowing the mixture to sit until cooled. This trick is probably not ideal (boiling kills most living organisms that might be in the water), but it's better than starving the birds that have become accustomed to your feeders.

Even if you live where hummingbirds stick around for winter or arrive early in the spring, don't change the four-to-one ratio of water to sugar. Where Anna's Hummingbirds remain year-round, some people advise enriching the mixture to three or three-and-half parts water to one part sugar to give the birds a little more boost in cold weather, but doing so risks causing dehydration in hummingbirds: during freezing weather, they need the water from the nectar in the absence of other sources, and a too-rich mixture doesn't provide enough hydration. My wintertime Anna's Hummers seem perfectly happy and healthy with a good, fresh four-to-one recipe.

Never use commercially available hummingbird food (the red-colored nectars sold in bottles and as sugary powder)—they contain colored dye, which is likely not healthy for tiny hummingbirds. And don't add red dye to your sugar water. Remember, red-colored food is not needed: hummingbirds are drawn to the color of the feeder; in fact, I suspect experienced hummingbirds could easily recognize a feeder by its shape and location, even if it had no red. Likewise don't add any sweeteners; use only plain white cane sugar, and avoid organic cane sugar because its natural molasses content has not been fully removed through processing.

Additionally, many hummingbird species are belligerently combative in defending their territories, and a single bird can dominate a feeder (or favorite patch of flowers), aggressively driving away competitors. The solution is to maintain two or more feeders out of line of sight from one another; that way, the birds can feed in peace at least some of the time. The same goes for flowers: if a single hummer routinely defends a favorite garden plot or even a single flowering plant, be sure to disperse other hummer-attracting flowers around the yard. And don't hang feeders in front of reflective windows—a hummingbird may mistake its reflection as a rival, attack, and strike the glass, possibly injuring itself. If you use a window-mount hummingbird feeder, add crossing or angled strips of colored tape to the outside window pane.

Most problems associated with hummingbird feeders are easy to manage and circumvent. You'll spend lots of time and effort keeping those feeders clean and the nectar inside them fresh, but it's critical that you do so for the birds themselves: keep 'em happy and keep 'em healthy.

Bees and Bats and Bears (Oh My!)

The rich, sugary liquid we put out for hummingbirds also appeals to a variety of insects, including bees, hornets, ants, and—gulp—yellowjackets, a type of wasp known for their aggressiveness and nasty, painful stings, which they seem quick to use at times. However, native bees, bumblebees, and honeybees are critical pollinators, and without

Hummingbirds seem unintimidated by bees, but the insects can make it impossible for the birds to feed. However, hummingbirds generally steer clear of yellowjackets and other aggressive wasps.

them, the ecosystem we rely on for our well-being would collapse. So only nonlethal methods should be used to keep these typically nonaggressive bee species away from hummingbird feeders.

Bees can certainly become problematic. I vividly recall the day I discovered that one of my hummingbird feeders was overrun with honey-bees—to the tune of perhaps 200 or more—happily devouring the nectar inside. I was amazed at how rapidly they drained it. I let them have their fill, while I looked up methods for keeping them away. I learned my problem had likely been exacerbated by the type of feeder itself—an inexpensive bottle-style model with no bee guards and with yellow highlights.

Bee guards are simple little screens that fit in the ports from which the hummingbirds feed, and they effectively prevent bees from being able to reach the sugar water. Many feeders come with them; for those that don't, you can buy packs of bee guards. As for the yellow color, well, it turns out that bees can't see red because their eyes don't have photoreceptors for it, but they see yellow remarkably well. Many bee guards are yellow—I'm not sure why, but I suppose they are made that way to contrast with the feeders, which are frequently red. Bees also have an advanced sense of smell: whether by vision or scent or some combination of the two, they often find hummingbird feeders. But they also give up if they can't reach the nectar. Some of my red-colored feeders have yellow bee guards, and I find them unproblematic, so if yellow attracts bees, they soon learn the pursuit of nectar at those feeders is unproductive.

The yellow plastic bee guards on this feeder plug into the ports and prevent bees from reaching the sugar water.

Feeder style can also eliminate bees. Besides being easy to use and clean, saucer-style feeders largely eliminate bees. The domed lid sits far enough above the sugar water in the saucer so that bees cannot reach the liquid.

While bees are generally easy to deal with by adding bee guards and/or choosing saucer-style feeders, yellow-jackets and hornets can be quite troublesome. Moreover, while I've watched hummingbirds feed in unison at a feeder being used by one or a few bees, the birds seem to be extra wary of even one yellowjacket or hornet: a nasty sting from one of these insects could be severely injurious if not fatal to a hummingbird. If yellowjackets or hornets become problematic, thoroughly clean your hummingbird feeders, being especially vigilant that no sugar water is spilled on the exterior of the feeders, and then move them. The hummingbirds will find them in very short order. Where I live, yellowjackets can be especially populous in some summers, and at times I have little choice but to hang yellowjacket traps at various locations around the property—not only to protect the hummers, but to make life outside bearable for dogs, family, and friends.

Ants, especially small sugar ants, as they are known colloquially, also love sugar water and can swarm feeders in unbelievable numbers. Saucer-style feeders generally have built-in ant moats, which, when kept full of plain water, effectively prevent ants from getting to the feeding ports. If you prefer bottle-style feeders, you can buy ant moats and traps that attach to the feeder's hanging apparatus and effectively eliminate the problem. Ant moats need to be monitored and refilled with water as needed; if ants swarm a moat, they can create a floating bridge of their drowned and drowning brethren, which provides egress to the feeder for the survivors, which will be legion. Also, earwigs seem capable of not only swimming the moat but then crawling into the ports and inevitably drowning within the sugar water, and the only solution is to keep that food fresh, changing it out often before dead insects can accumulate in it.

ABOVE Sugar ants got to this feeder and easily crawled through the feeding ports, swarming the nectar inside and leaving many of their number drowned.

BELOW Even large carpenter ants are attracted to sugar water; these ants crawled down the line suspending this saucer-style feeder and, with the ant mote dry, were able to crawl into the ports.

Although sometimes touted as an option to deter ants, smearing petroleum jelly or other such solutions on the hanging pole or suspension lines of your feeders risks getting those substances in the feathers of birds (not to mention, it's a messy proposition, especially in hot weather). Also, don't use ant guards/traps that contain insecticides—they may be harmless to some species, but not all. Besides, water-filled ant moats frequently serve as a source of drinking water for small songbirds—mine are routinely used by chickadees, nuthatches, finches, and warblers.

Beware too of praying mantises, which are known hummingbird predators, rare though such instances may be; if you find a praying mantis on or near your hummingbird feeder, gently move the fascinating insect far away to a new location. There is no perfect solution to keeping unwanted insect visitors away from your hummingbird feeders, but with appropriate products, monitoring and maintenance, and moving the feeders as needed, you can eliminate serious problems.

And then there are the furry problems that occur, albeit rarely, in some areas. House cats and their feral brethren are only too happy to exercise their predatory instincts trying to catch hummingbirds, and they occasionally succeed. Deter cats by leaving them no ambush points near your feeders—make sure the feeders are well clear of concealing brush, shrubs, and structures. If you have outdoor cats, consider skipping the hummingbird feeders altogether.

Yummy sweet sugar water is like candy to black bears, raccoons, squirrels, and, in the Desert Southwest and southward through Central

Raccoons, as well as squirrels, flying squirrels, coatis, nectar-eating bats, and even bears will help themselves to the sugary snack provided by a hummingbird feeder.

America, coatis. In many cases, these mammals get to the sugar water in ways that are not exactly healthy for the feeder itself. The only solution short of removing the mammal—and its kind were probably there before our kind—is removing the temptation. If bears or raccoons or other mammals decimate your hummingbird feeder, move it, perhaps to a higher location, suspended in a manner that makes it inaccessible to the furred marauder with a sweet tooth. Or move it to a window mount. In extreme cases, you may need to desist with the artificial feeders and count on your flowers to attract hummingbirds.

In many places—such as the Desert Southwest, Mexico, and Central America—hummingbird enthusiasts wake in the morning to find their feeders emptied but undamaged. This is typically the work of nectar-feeding bats, which evolved to feed on the nectar offered by flowers. Like hummingbirds, they easily make the jump from flowers to feeders. Bats are wonderful creatures to have around your yard, so if these nocturnal visitors are draining your hummingbird feeders, consider yourself fortunate: you are now a hummingbird feeder and a bat feeder.

A variety of birds beyond hummingbirds are also drawn to sugar water, including woodpeckers, orioles, tanagers, warblers, vireos, and others. Watching them try to perch on hummingbird feeders in a manner that allows them to slide their tongue into the ports can be entertaining, and such visitors are rarely more than an inconvenience to the hummingbirds.

This Gila Woodpecker and his fellows are reasonably adept at drinking sugar water from hummingbird feeders.

Wintry Weather

All hummingbirds found in the United States are migratory and may pass through your area or arrive at their summer breeding grounds in the spring, long before summery weather sets in. Not only must they feed themselves before flowers are blooming in abundance, but they must also contend with cold and wet weather. They are supremely adapted to doing so, having evolved their migration routes eons ago. These northern species rely heavily on insects for food, but they need to fuel their high-speed pursuit of those

In the Pacific Northwest, Anna's Hummingbirds are now year-round residents.

insects, so they feed on available flowers and even tree sap during migration. They also benefit from hummingbird feeders.

Early-arriving hummingbirds, along with species that live in the mountains, must also contend with subfreezing temperatures at night, and they do so by entering a state of torpor, a radical slowing of the metabolic rate. The Anna's Hummingbird, in fact, having rapidly expanded its range northward, doesn't bother migrating south from its relatively new home range in the Pacific Northwest—and in staying put for the winter, aptly demonstrates the toughness of these little flying jewels.

But at times, hummingbirds dealing with inclement or cold weather can benefit from a little helping hand. Hummingbirds don't mind water, but a cold shower on a cold day isn't much fun. A simple weather guard attached above your hummingbird feeder provides dry space for them to feed. Commercially made canopies are readily available, or you can make your own with something as simple as a red plastic picnic plate. Snow and ice are more troublesome, and Pacific Northwest hummingbird aficionados have learned to deal with these wintertime hassles thanks to the year-round Anna's Hummers.

In freezing temperatures, hummingbirds awakening from torpor in the morning need to replenish their energy very soon; and they need to feed just before roosting for the night. In the dead of winter, food sources are generally limited to tiny insects and sugar water. But if that sugar water freezes overnight, the birds can quickly become imperiled from lack of fuel, and tales of winter hummingbird rescues for such reasons are commonplace, especially in the Pacific Northwest range of the Anna's Hummingbird. So it's imperative to prevent feeders from freezing. The easiest way is to bring them inside at night when the forecast calls for subfreezing temperatures overnight, but doing so begets the responsibility of putting the feeders back outside at the crack of dawn if not before: I am constantly amazed that Anna's Hummingbirds are routinely active before the sun crests the horizon even on the coldest days in January. Moreover, you must wait until after dark to bring the feeders indoors to assure that the hummers have a chance to fuel up before they retire for the evening.

The only other alternative is to find a method of preventing the sugar water from freezing when the feeders are left outside. Placement can help: choose a location sheltered from the wind. If you hang the feeders up against a sun-exposed south face of a wall, they'll gather a bit more heat than if left out in the open. You can also hang them under a porch roof or overhang; I hang a feeder under my carport during the summer, so the birds are already accustomed to using it.

When placement alone isn't effective in combating the cold, preventing sugar water from freezing requires ambient heat. Northwesterners, in particular, have come up with a variety of innovative heat sources; the easiest, safest, most reliable antifreeze method I've tried is the Hummer Hearth (hummerhearth.com), invented by a hummingbird fan in Washington state. The Hummer Hearth uses a 7-watt candelabra C7 light bulb to provide just enough heat to prevent sugar water from freezing, down to about 15 degrees; it is designed to fit feeders equipped with a ring-style perch and attaches to the bottom of the feeder. The unit is powered by a 6-foot electrical cord, so you need to place it within reach of an outdoor outlet and likely you'll need an outdoor-rated extension cord. You can swap out the 7-watt bulb with a 15-watt bulb to prevent freeze-up at even lower temperatures.

A variety of other methods are also effective against freeze-ups. In the realm of plug-in electrical options, you can keep feeders warm with plumber's heat tape or heat cables, or by hanging a light from or near a feeder. Heat tape and heat cables are rated for use in wet conditions (they are used to prevent pipes from freezing). But beware that many types of heat-producing light fixtures are intended for indoor use and can fail, sometimes dangerously, if they are exposed to moisture. That's

Convenient and easy to use, the Hummer Hearth uses a low-watt bulb to provide enough heat to prevent sugar water from freezing in low temperatures.

why I don't recommend using electrical lights such as trouble lights (aka drop lights, inspection lamps) and hooded lamps (aka clamp lights, heat lamps) unless your hummingbird feeder is in a location fully protected from weather.

Other proven solutions include attaching a bundled string or two of outdoor-rated Christmas lights, with incandescent (not LED) bulbs, to the underside of saucer-style feeders or wrapped densely around bottle-style feeders; and using duct tape or rubber bands to attach chemical hand- or body-warmers to feeders. Be sure to buy the longest-lasting models (such as HotHands). The hand-warmer method is handy if you don't have an exterior electrical outlet.

During one particularly nasty cold spell in western Oregon some years ago, I was acutely worried for the survival of three Anna's Hummingbirds that had remained around my yard for the winter. I was leaving town for three days, and overnight temperatures were expected to dip into the low 20s that week. My house didn't have any external power outlets, so I couldn't run an electrical cord for a light or for heat tape, and I'd be gone too long for handwarmers to be of much help. So I insulated the feeders as best I could, enshrouding them in dense foam insulation taped into place so that only the feeding ports were exposed. Already my feeders were placed in the sunniest available locations. I worried about the birds all through the trip but came back to find them active and apparently unfazed.

That same winter, my sister found a male Anna's Hummingbird belly-up on the porch below her hummingbird feeder. She brought the little fellow inside, found him to be lethargic but alive, swaddled him in soft cloth in a bucket, and put a screen over the top. She then filled an eyedropper full of sugar water and was able to feed him by touching the tip of his bill with the little bubble of nectar dripping out of the eyedropper. She repeated the feeding a few times over the course of 45 minutes or so; soon he was back to normal, and she set him free. Upon inspecting her feeder, my sister discovered that while the sugar water was still liquid, the feeding ports were completely frozen over and plugged with ice—the bird simply could not get to the food.

Therein was a valuable lesson: check those feeding ports, even if the sugar water is still liquid. Use a toothpick to probe the ports to check for and clean out any ice; during snow, sleet, or freezing rain, actively monitor your feeders and clear the feeding ports as needed; and consider adding a canopy to protect the feeders from precipitation. With diligent helping hands from hummingbird fans, those nonmigratory Anna's Hummers, as well as winter vagrants and early-arriving migrants throughout the United States, do just fine during subfreezing weather. They are indeed tough little mighty mites.

THREE

Planting & Landscaping for Hummingbirds

If you love hummingbirds but don't have a proverbial green thumb, don't sweat it: neither do I. Successfully gardening for hummingbirds is easy and a great way to get children outdoors and immersed in nature. In my earliest days of gardening, I began to think I could kill just about anything I put in the ground. But soon I realized that growing flowers is really just a matter of choosing climate-appropriate plants, placing them where they get the right amount of sunlight, giving them good soil, and watering enough but not too much. You needn't be a master gardener to establish flowers that draw hummingbirds, and the reward for even a modest effort is hours and days and years of entertainment and awestruck wonder as the meteoric mini-birds enjoy the fruits of your labor.

My first backyard overhaul aimed at creating a hummingbird-friendly landscape began with some fits and starts. I built a rock-contained raised garden, anchored at one end by torch lilies (aka red hot pokers), with tall-growing trumpet lilies behind them along the fence, columbines in front of the red hot pokers, and Cascade Blue rock

Rufous Hummingbird.

cress (*Aubrieta*) growing along and down the rock faces. At the other end of this raised plot, I went with beebalm along the fence and Cape fuchsia in front, with more of the rock cress to drape the stones I had used as a low enclosing wall.

Everything worked out as planned for a couple of years: the rock cress and columbines bloomed early, providing a food source in spring when the hummingbirds arrived; the red hot pokers opened as the columbines faded and lasted well into summer; then the lilies, 4 feet tall along the fence, flowered, lasting into late summer. At the other end, the Cape fuchsias did well the first year, blooming by early summer, and even better the second year, sending out trailers and new sprouts; likewise, the beebalms quickly shot up along the fence and began blooming in midsummer, lasting into fall. The hummingbirds loved the garden.

But just as I was patting myself on the back for a job well done, things started going sideways: the red hot pokers decided to be the big bosses. In year three, they exploded, their basal tufts reaching more than a foot across, and their strappy leaves and gorgeous flower spikes towered over and obscured the sun-loving lilies. Their profusion also threatened the columbines, which had shown excellent self-sowing progress—until being bullied for sun and space by the red hot pokers. The denizens of the other end of the plot were better behaved and acted as I had hoped—until the Cape fuchsias shaded out the rock cress. I had work to do. I let the red hot pokers enjoy their ascendancy until fall, and then I divided the clumps and removed about half of each. I pruned the Cape fuchsias to allow ample sunlight to reach the rock cress.

Equilibrium returned, but only with fastidious attention and garden management. It was a valuable lesson: successfully maintaining any garden takes foresight and work, but when you plan your gardens specifically to attract hummingbirds, you must often work just a little harder to keep the little critters happy and healthy.

That said, hummingbirds are accommodating, and if you don't have the time, space, or desire to establish and maintain a large garden or entire yard dedicated to attracting the bejeweled little birds, you can still enjoy their company with a minimum investment. Two or three potted flowers, a couple of hanging baskets, a single flower bed, or just the right kind of flowering vine on a trellis or arbor might be all you need to attract hummingbirds.

Not all flowers are equal in their ability to attract hummingbirds. The birds need nectar, and when they find a favorite food source—whether an entire yard, a garden plot, or even a single flowering plant—they tend to develop site fidelity. They'll return day after day and year after year, especially if they nest in the area.

Brightly colored tubular flowers are best, and while hummingbirds seem especially drawn to shades of red, orange, and pink, they readily feed at flowers of any color that offer them high-quality nectar. An abundance of red and orange blooms does seem to draw their attention, and once attracted to your yard, hummingbirds

Establishing a hummingbird garden is easy: choose flowers appropriate to your climate, plant them in locations that meet their light and soil needs, water as needed—and watch them flourish.

will investigate flowers of all different colors and quickly identify their favorites.

Remember that in addition to flowers for nectar, hummingbirds need secure shelter within trees and shrubs for nesting and overnight roosting and prominent daytime perches from which they can survey their territories. Many species of hummers are adapted to prefer sunny openings in otherwise wooded areas—the trees and shrubbery provide nesting and roosting habitat, while the openings, such as meadows and swales, provide prime conditions for many different flowering plants. Hummingbirds also need water; they preen meticulously, happily using a bit of water to help, and they also drink water, though they can meet their water needs through the nectar they eat. They especially like mist, but dripping water and water features also attract them.

So, an ideal hummingbird yard includes all these elements—trees and dense shrubbery, open space, plenty of nectar-rich flowers, and water. Provide for their needs in abundance, and you are likely to routinely host hummingbirds; provide for their needs in a minimalist arrangement, and you still stand a good chance of attracting them, at least in transit, and maybe even more permanently if the general area in which you live offers everything the birds need.

Designing and Establishing a Hummer Garden

Hummingbird-friendly yards and gardens begin with well-conceived plans that account for the hummingbird requirements of food, water, and shelter. If your affinity for hummingbirds leads to extensive renovation of your yard, begin with a drawing so you can better visualize your ideas. Sketch your yard more or less to scale and then decide what elements you want to add: narrow gardens along fencerows or exterior walls, raised flowerbeds, containers, trellises or latticework, shaded plots and sunny plots (which might require planting or removing trees or large shrubs), lawn borders, and maybe water features. Think vertically, not just horizontally: if you have space or can create space for climbing vines and tall-growing flowers, these vertical layers add substantially to the available food sources you can offer to hummingbirds.

This backyard hummingbird garden takes full advantage of limited space by layering flowers vertically.

Next, decide which plants will go where, and write these onto the blueprint. The bulk of this chapter identifies many of the best flowers for attracting hummingbirds, but in choosing

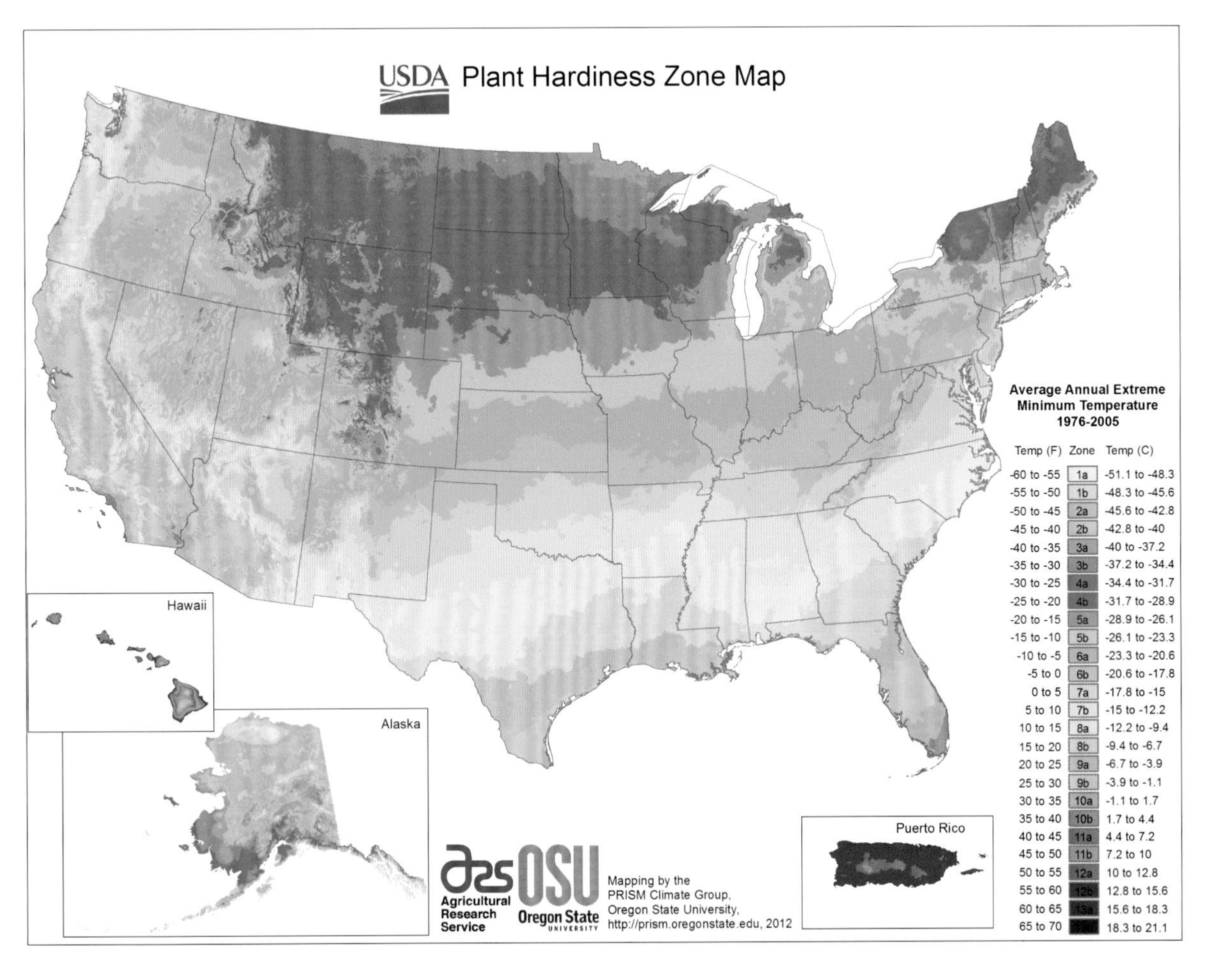

The easy-to-use USDA Plant Hardiness Zone Map is your guide to choosing flowers that will thrive in your area.

your favorites you must consider the habitat and climate preferences, and growth and blooming behaviors, of each plant species. To learn which hummingbird-attracting flowers will thrive where you live, consult your favorite local nursery and the USDA's Plant Hardiness Zone Map (planthardiness.ars.usda.gov/PHZMWeb), where finding your climate zone is as easy as typing your zip code into the box provided.

Hardiness zones are important. A plant rated for zone 8 and higher is not likely to do well, or even survive, in zones below that on the scale (and likewise, many plants well adapted to cold climates don't do well in places that are too hot for them). But finding your climate zone is only half of the equation. The other is finding plants that match your climate zone. Climate zones are provided for each of the hummingbird-attracting flowers

Beebalm is hardy down to zone 4 (3 with protection), making it an excellent choice in cold-winter regions. Here it is mixed with crocosmias (the swordlike leaves), which are hardy to zone 6.

described in this chapter, but for the myriad other great hummer flowers available to gardeners, you need to match their climate zone preference to the zone you live in. A simple online search will reveal a plant's hardiness zones via any number of different gardening and nursery websites.

Now all you need to do is source the plants, ideally by paying a visit to your favorite local nursery; the small identification/information tags that routinely accompany plants at retail outlets usually list the plant's hardiness zone. You can hedge your bets by choosing plants that are rated at least one zone higher and lower to guard against extremes in weather. Also consider using native species as much as possible, and avoid allowing non-native species to spread into wild areas. Native species often produce more nectar than intensely cultivated flowers selectively bred for large, colorful blossoms, explaining why some cultivated varieties (cultivars), even though they may look like good hummingbird flowers, fail to attract the birds, especially if better nectar sources are readily available. The Lady Bird Johnson Wildflower Center (wildflower.org) is a great source of information on native species.

A garden renovation can be anything from a major makeover to modest additions. The most elaborate flower gardens include decorative elements, irrigation systems and water features, and fastidious maintenance—but such plots are the realm of professional landscapers and horticulturalists. We don't need to recreate the Royal Botanic

This astutely designed garden uses a variety of hummingbird favorites and also provides water, nearby perching sites, and plenty of open space.

Gardens; we just want to make a yard, big or small, attractive to hummingbirds so we can enjoy their company. That might mean simply adding a few hummer-friendly plants to your existing gardens, yard, or patio, or it might mean a full overhaul of your backyard. Much depends on how much of an investment of time, labor, and money you want to make—and on how dedicated you are to investing in hummingbirds. I've managed both ends of the spectrum, having redesigned an entire backyard (albeit a small one) at one house, and at a later residence, using a less-invasive, less labor-intensive, and less costly approach by relying on a few select ground plantings and lots of container plants. Both cases are instructional.

The small backyard I redesigned began life as the woeful remnants of the lawn laid down by the contractors who built the house. Unfortunately, the drainage pattern in the area, though subtle, ran through my backyard, so during the rainy season (from mid-autumn through spring in the temperate zone where I live) the would-be lawn was a swamp. With each passing year it fostered more and more moss; even ferns popped up. During the warm summers, however, that swamp became a lawnlike fallow atop clay-based soil that, baked dry by the sun, was so hard it was barely penetrable. If left unchecked, the remnants of the original lawn grass, joined by a litany of weeds, grew rapidly, requiring that I routinely mow it, which in turn rendered it parched by the summer sun. Eventually I decided to do something about it, first drawing out a rough sketch of a backyard that featured narrow, rock-walled raised gardens around the edges and two centerpiece raised gardens, with 2-inch round river rock filling in all the spaces between the low rock walls.

The plan was labor-intensive; it needed to be, for this was a landscaping job, not just a case of growing some new flowers. Because the ground was too hard to dig in during summer, I waited until March, and then began by lining off the approximate locations of the raised beds. I dug them out to a depth of about a foot to loosen that native clay

I overhauled my backyard with an eye toward planting for hummingbirds. The first step was the most labor-intensive: building raised beds and river-rock pathways.

soil, and then with a square-point shovel, leveled the areas outside where the flower beds would be and deposited that soil, complete with organic material, into the plots. Next, I built the rock walls from granite blocks available for bulk purchase by weight from a local landscaping center. Once the walls were in, I lined their inside edges with landscaping fabric so that, later, when I watered the plants, muddy water would not leak out between the stones. I added high-quality garden soil and kept it wet so it would compress. Finally, I anchored heavy ground fabric in all the spaces between gardens and filled those places with river rock, forming pathways around and between the garden plots.

In one of the centerpiece gardens and in the garden along the back fence, I also installed landscaping fabric because my plan for those areas relied on plants that would not spread much underground. In the other gardens I did not use fabric because I intended to plant a variety of flowers that would self-sow or spread by underground taproots. The tradeoff, of course, would be more effort to control weeds in the gardens without fabric. Once the fabric was pinned down, I cut holes for each plant and, using starts from a local nursery, dropped the plants into place. Once the plants were properly embedded, I covered the fabric with a mulching bark dust and then planted the gardens that did not have fabric, likewise adding a mulch layer.

The plants were in the ground. For that first spring, I included some annuals that would bloom quickly and a variety of perennials. Some of the perennials would bloom early in the season and others later, but the primary goal was to get them established with an eye toward the future—perennials are just that: plants that live for multiple years. I could worry about adjusting my mix of plant species as the gardens became established. Seeking verticality (and privacy), I added trellises

Once the flowers beds were ready, I mixed annuals with perennials, hoping the annuals would attract hummingbirds while the perennials were establishing themselves.

for the flowering vines I had planted at the side fence posts and a large screening trellis with flowering vines along two sides of the small patio; seeking abundance, I spotted some potted flowers in strategic locations. Hummingbird feeders on shepherd's hooks completed the redesign.

I made mistakes; some design elements didn't look or behave as I had hoped, and some plants proved troublesome, but nature is forgiving (especially in temperate climates), and the gardens soon flourished. The hummingbirds visited often, increasingly so by the second and third years, when the flowering attractants I had included—beebalm, Cape fuchsia, penstemons, red hot pokers, and honeysuckle—were well established and bloomed profusely.

By its third year, my backyard was well on its way to becoming a hummingbird hangout.

Some years later, at a different residence, I was able to draw on my experience to create a hummingbird space without redesigning the entire yard. The rented house was bordered on one side by river bottom that had been used intermittently as pasture and which flooded each winter. The property owners had established a flat border around the house, which sat above the winter flood level, and covered most of that area with pea gravel atop landscape fabric. The pea gravel extended around one side of the house; the other side was a driveway. Beyond the 20-foot-wide flat backyard, the ground fell steeply away to the floodplain, about 10 vertical feet below. The far side abutted a steep hill enveloped in impenetrable briars.

Immediately upon moving there I set up hummingbird feeders, taking a cue from the neighbor, who had three feeders that seemed always abuzz with hummers. But from the outset it was clear that most of my planting would be contained—to pots and other vessels. One short fence line needed some anchoring vegetation to retard erosion down the slope behind it—no coincidence that I chose favorite hummingbird shrubs to do the job, species I'd had great success with previously: Cape fuchsia and Hot Lips salvia. I bought well-established plants from a favorite local nursery, placed them in the ground in May, and the hummingbirds found the flowers before my shovel was dry. Because the hummingbirds had long since begun frequenting the two feeders within eyesight of that corner of the fence line, they were no doubt keen to check out any bright new flowers in the vicinity.

Flowering container plants can add layers and serve as hummingbird magnets wherever they are placed. Here, a porch staircase is space enough for some potted columbines and other hummingbird-attracting flowers.

But beyond that, I supported the effort with potted plants. I chose flowers that hummingbirds like and tried to create a mix of species so the birds would have nectar sources all summer. As always, the hard part, or rather the time-consuming part, was maintenance, but watering, fertilizing, pruning, deadheading, and the like are part of the fun. The hummingbirds seemed happy (at least in terms of the food supply)—if happiness can be judged by constant fighting over flowers and feeders!

Creating a Hummingbird Haven

Some hummingbirds that arrive in spring have reached their favorite territory; others may simply be passing through your yard on the way to breeding sites farther north (or east or west). At the end of the nesting season in late summer, in many locations, modest populations of breeding hummingbirds are bolstered by southbound migrants. So, in selecting plants to feed hummers, make sure you have spring-blooming flowers as well as late-summer/fall-blooming flowers. By doing so, you help the migrants fuel up during their lengthy journeys, and you stand a better chance of making your yard a stopover location for years to come. Moreover, an abundant and varied supply of flowers may reduce a hummingbird's dependence on sugar-water feeders—flower nectar has vital nutrients lacking in sugar water.

Staggered blooming periods among your different flower species assures hummingbirds a season-long supply of food, so choose plants based on both their attractiveness to hummers and their bloom time. Perennials form the core of a hummingbird garden, with annuals making valuable additions because they can help fill in during times of the season when certain perennials fade and others are just beginning to bloom. Annuals can also diversify your hummingbird offerings. Through well-conceived garden design, you can create successional blooming periods among myriad flowers, and this concept is key to creating ideal hummingbird habitat.

Keep plant size in mind, and layer flowers vertically, so that the tallest plants are in the rear and the shortest ones are in front. Especially in border areas—fences, hedgerows, exterior walls—take

LEFT Hanging baskets of fuchsias and other hummingbird favorites are great additions to any hummingbird garden and also ideal stand-alone hummingbird magnets where space is limited.

BELOW Although tiny, this thoughtfully designed space relies on potted plants, arranged low to high, to draw hummingbirds.

the advice of *Southern Gardening* podcaster Gary Bachman: plant low to high; not only does this "stadium staging" look great, but it also provides easy access for hummingbirds and lets you, like baseball legend Bob Uecker, "sit in the front row to watch the action."

In center areas, follow that same basic concept, establishing the tallest plants in the middle, and the lowest-growing flowers on the periphery. This concept holds for in-ground garden plots and containers. Use elevation in other ways also: add hanging baskets with favorite hummingbird flowers; build stair-stepped gardens big or small; add tall trellises or arbors for flowering vines. Filling some vertical space allows you to add even more flowers to the array that attracts hummers to your yard.

Also try to anticipate your garden's evolution: some plants spread readily, either by underground rootstock or by self-sowing, so plan for such eventualities in your selection of flowers and in how you manage them once they are established. Large drifts (groupings) of one kind of flower provide lots of nectar per square foot for hummingbirds, and such massed flowers make great visual elements in the garden, but you must manage their growth, not only so they flourish, but also so they don't overwhelm nearby plants.

Use only organic fertilizers and mulches; many chemical agents are potentially harmful to hummingbirds. Avoid insecticides; if you must use them to save certain plants from being annihilated, opt for ecofriendly and pest-specific varieties.

Hummingbirds need protein and therefore eat lots of insects, so other than in extreme cases of depredation, it's generally best to let nature work things out in your garden with a helping managerial hand from you. Hands-on gardening—getting out there every day to examine and manage your plants—allows you to find and correct problems, such as insect damage, rodent issues, and powdery mildew on some plants, before they become acute.

If hummingbirds frequent your gardens and feeders through the summer, the birds are likely nesting in the area, so be sure to leave any spiders you see at this time to their own devices and their webs intact: spider webs are a crucial binding material for hummer nests. And of course, spiders also prey on insects, many of which are plant damagers.

Provide water alongside your nectar-rich flowers, and your yard may well become a busy hummingbird central station. These tiny birds love a good shower, and a misting hose is a great way to provide them the means. In fact, you can condition hummingbirds to routinely use a mister or dripping water source by using a hose timer set for the same times each day—inexpensive timers are available at hardware and home supply stores. Hummers will also bathe in and drink from standing water in a shallow bird bath; keep it shaded as much as possible, so the water remains cool, and refresh it each morning and evening when you water flowers. Add a few flat-profile stones that stand just above and just below the water's surface.

BEST BETS FOR HUMMINGBIRD BLISS

THROUGHOUT THEIR RANGE, nearly from pole to pole if you consider Rufous Hummers in Alaska and Green-backed Firecrowns in Tierra del Fuego, hummingbirds feed on nectar from thousands of different flowers. The choices for hummingbird-inspired gardeners are legion, and it's hard to go wrong if you keep the basics in mind: hummingbirds need nectar-rich blooms, and they prefer tubular flowers in shades of orange and red (although yellows, pinks, and purples attract them, too). Beyond that, as you've just heard, you must match flowers to the climate zones in which they thrive and then tend them as needed. Among the thousands of flowering plant species utilized by hummingbirds, hundreds are available through gardening retailers, ranging from large chain stores to locally owned nurseries to regional and national outlets specializing in native plants. Some of the best of those hundreds of ideal plants are listed here, a mix of native and exotic species.

BEEBALMS (*Monarda*)
Zones 4–9

A member of the mint family and a North American native, beebalm (aka horsemint, Oswego tea) is a fragrant favorite of bees, butterflies, and hummingbirds. The leaves, which were traditionally used both medicinally and for tea, have a musky, spicy scent, and the tubular flowers grow densely from bulbous flower heads atop stalks that can reach 5 feet tall, depending on species and growing conditions. In some species, the flower clusters are spectacularly showy. Native *Monarda* species are variously colored, from white and pale yellow to deep red to purplish. One of the best is *M. didyma* (scarlet beebalm), which is native to Oregon and Washington and widespread (but not necessarily common) in the eastern United States and eastern Canada. This species, available through nurseries, makes an excellent anchor to a hummingbird/pollinator garden because, like other *Monarda* species, it has a lengthy blooming season, often from early summer into early autumn—not to mention being an extremely strong attractant for hummingbirds. The bright red flowers are beautiful.

Monarda fistulosa (wild bergamot) is another native, whose range virtually spans the continent in both directions. Its flowers range from light pink to bright fuchsia. About 15 other species round out the genus, varying in their relative attractiveness to hummingbirds, but you can't go wrong with *M. didyma*, *M. fistulosa*, or any one of the dozens of cultivars that are bright red, crimson, or magenta.

Beebalm grows best in sunny locations with consistent watering; it thrives in rich, loamy soils high in organic content but does well in just about any well-drained soil. Plant starts about 18 inches apart; the plants will spread via underground stolons but are easy to thin and contain. Beebalm doesn't like to dry out completely, so a layer of mulch or compost around the established plants helps retain moisture. Once established, beebalm spreads to form dense colonies. It makes an excellent back border for stair-stepped gardens because it grows tall and blooms at the top of each stem; plant shorter flowering plants in front of it. Beebalm also makes great bookends—plant them at the terminal ends of a raised garden or in the sunny corners of a fence line. The individual stalks need space for good air circulation because they are prone to develop mildew on the leaves (though some cultivars are resistant to this problem). Deadhead the spent blooms to encourage regrowth, and after the first fall frost, cut the stems back to just a few inches tall. Beebalm is hardy to zone 4, but with good winter care (a layer of mulch over the plants when they've been pruned back), it can thrive in zone 3.

ABOVE Beebalms are very attractive to hummingbirds (and bees and butterflies).

BELOW Beebalms make an excellent anchor for an entire hummingbird garden.

BLAZING STARS (*Liatris*) Zones 3–9

Liatris is a large and widespread genus of native perennials, most of them sporting spirelike tufts of pink or lavender flowers, held like a bottle brush atop tall, rigid stems that emerge from a basal rosette of leaves. The showy blossoms of blazing stars (aka gayfeathers) attract hummingbirds, bees, and butterflies and have the added advantage of reaching peak bloom in mid- to late summer. *Liatris spicata* (dense blazing star), *L. pycnostachya* (prairie blazing star), and *L. punctata* (dotted blazing star) are the species most commonly available for gardeners, alongside several excellent cultivars. These and most *Liatris* species are native to the eastern half of North America, especially the Plains states, and several varieties do well in cold climates, making them especially valuable for hummingbird (and butterfly/bee) enthusiasts in the north.

Blazing stars are best established from plant starts or corms, which are often sold as "bulbs." Corms are storage organs that grow on a plant's underground stem base. Select the largest corms available; in spring, plant them 2 to 3 inches beneath the surface in well-drained soil, about 18 inches apart, and water thoroughly; avoid sites that collect standing water. When the plants emerge, they need regular but light watering—they are subject to root rot with overwatering. To guard against frost damage in northern climes, layer a few inches of mulch over the plant mounds in autumn.

Blazing stars like full sun, especially in northern latitudes, and tolerate drought. Because they can reach 4 feet in height (some cultivars are even taller), blazing stars make great endpieces, stand-alone plants, and fence liners, adding beauty

ABOVE Native blazing stars thrive at Big Oaks National Wildlife Refuge in Indiana.

OPPOSITE *Liatris* is a genus of hardy wildflowers that hummingbirds love.

to any garden. They grow well in pots and also tend to mix in fine with other flowers, allowing you to create multispecies plots that serve as magnets to hummingbirds.

In my first attempt with blazing stars, I planted three of them too close together; nevertheless, they interacted fine with early-blooming columbines planted all around them and all-summer-blooming Cape fuchsia. The columbines needed water throughout the same weeks my young liatris were sprouting and needed daily water during summer; in their third year, my blazing stars were so robust that I divided them (carefully) by shovel in the fall.

BUTTERFLY WEED
(*Asclepias tuberosa*)
Zones 3–9

Native to most of the United States and eastern Canada, butterfly weed (aka butterfly milkweed, orange milkweed) is a shrubby herbaceous perennial that produces tightly clustered orange to reddish blooms. Not only is it beautiful, it is tough, able to withstand severely cold winters, making it a staple in hummingbird gardens in northern climates. The mature flower stalks, 12 to 30 inches in height, bloom from early summer until the first frost (depending on location); that means butterfly weed is a valuable seasonlong food source for hummingbirds, not to mention butterflies and bees.

Another plus: butterfly weed is happy to occupy the less desirable parts of a garden—areas with poorer soil, too much drainage, or a bit too much sun. It also thrives in pots and planters and makes a great accent for patios, decks, and walkways. Provide the plants with plenty of sunlight.

Butterfly weed forms seed pods, and the seeds are dispersed by wind; under ideal conditions, they

ABOVE Hardy and resilient, butterfly weed is an especially welcome addition to any pollinator garden.

OPPOSITE Butterfly weed is a great choice for hummingbird gardens, especially in northern climes.

take readily, so if you don't want these plants establishing themselves at random locations, remove the seed pods as they form. Plant seeds in the fall; they need the overwinter cold to germinate and sprout in the spring. New plants generally flower in their second year (water them regularly the first year). After the first frost or in early spring, cut back the stalks; if you want the plants to self-sow from their seed pods, leave them intact until early spring. A layer of mulch over the plant's crown helps assure winter survival.

Butterfly weed is easy to establish from starts, cuttings, or—given its propensity for self-sowing—seeds. Because it spreads by underground rhizomes, consider containing butterfly weed in raised beds or pots; however, I've found it easy to control if I simply pay attention to where new shoots sprout up and remove them as needed.

CALIFORNIA FUCHSIA
(*Epilobium canum*)
Zones 7–10

I was awestruck the first time I saw California fuchsia (aka hummingbird trumpet, zauschneria) in its native habitat. On a fly-fishing expedition down the wilderness stretch of Oregon's Rogue River, I spotted dazzling accents of bright red along the basalt walls of Mule Creek Canyon. My guide told me what they were and acquiesced to my request to beach the drift boat so I could examine the flowers more closely. They grew from crevices in the rock, where just a bare minimum of soil allowed purchase for their roots. Able to thrive in such an environment, the flowers are tough customers. If California fuchsia were better known among gardeners, and particularly hummingbird enthusiasts, it and its cultivars would no doubt be more prevalently available from nurseries.

California fuchsia is a clump-forming woody-based perennial that produces scarlet to reddish orange tubular flowers in abundance amid grayish green leaves. Rather than deteriorate on the stem as they fade, the blooms simply drop away, and the plant tends to bloom continuously from late summer well into autumn. In

ABOVE California fuchsia can form both bushy shrubs and creeping shrubs that drape over rocks, rails, and walls.

OPPOSITE In places where it thrives, California fuchsia is just about a lock to draw in hummingbirds.

warm, relatively dry areas they are easy to grow provided you don't fuss over them too much: they are adapted to rocky ground, so in domestication all they really need is well-drained soil and sun. They prefer slopes, making them excellent accents to stone garden walls, particularly the cultivars developed for use as groundcovers, such as Everett's Choice and Orange Carpet (the latter prefers a bit of shade and a tad more water than the typical drought-tolerant varieties). Water new plants daily; once established, the plants need only minimal water, and hummingbirds find them irresistible.

An online search will lead to retail outlets that carry starts of this native plant. Plants can also be grown from cuttings, and large plants can be divided in fall or winter into stems with roots and rhizomes for replanting in a new location. Leave the seed pods on the stems in fall to allow the seeds to disperse naturally and self-sow. Leave the spent stalks uncut for winter and clean them up when new growth appears in spring.

CAPE FUCHSIAS (*Phygelius*) Zones 7–9

Despite their common name, Cape fuchsias (aka Cape figworts) are not closely related to fuchsias (which are also hummingbird attractors). The two species—*Phygelius capensis* and *P. aequalis*—are native to southern Africa and have given rise to myriad cultivars, most of which do well in temperate and even cool climates.

These gorgeous tubular flowers grow profusely on rigid stems, and hummingbirds love them. Use the red, magenta, coral, and other reddish shades to draw in hummers, but you can intersperse them with the yellow varieties (aka yellow trumpets) to add contrast to your gardens. The hummingbirds will readily feed from all colors. Whereas the more familiar fuchsias are frequently planted in hanging pots, Cape fuchsias are in-ground shrubs that thrive in well-lit areas; they can also be contained in planters. They send out underground creepers and spread rapidly once established, sometimes growing more than 4 feet tall and wide. With enough time, given good growing conditions, they can fill out sections of garden. If this is a concern, plant them in large plots or just meticulously prune them. Cape fuchsias are great for rock-walled raised beds because the creepers will reach the gaps in the rocks and sprout through them.

Plant Cape fuchsias in well-drained soils and water them consistently. They do well as stand-alone plants and often thrive when established along fences and exterior walls, where they can bask in the morning sun. They will withstand full sun, especially in zones 7 and 8, but water them regularly and be wary of too little or too much

ABOVE A variety of Cape fuchsia cultivars are available.

OPPOSITE Cape fuchsias, particularly the red-flowering varieties, are winners for hummingbirds.

water: Cape fuchsias like to remain just a little moist but not wet. Discolored and dying leaves can indicate too much or not enough water (or can indicate certain pathogens that can affect this plant, in which case, consult a local nursery, horticulturalist, or university extension service).

Cape fuchsias bloom all summer, and better still, as each stalk of flowers dies, you can deadhead them by clipping the stem below the spent blooms at the next joint that shows sprouting leaves. These plants respond well to such pruning and will continue to produce new flower stalks throughout summer and into the fall. With such care, all-summer-blooming Cape fuchsias serve as great anchors for your hummingbird-friendly garden; healthy plants tend to fill out robustly by their third year. Arrange them with early- and late-blooming hummer favorites, as well as flowering groundcovers and tall, slender hummingbird favorites, such as beebalm. Beware, though: deer eagerly graze away new growth on Cape fuchsias.

CARDINAL FLOWER
(*Lobelia cardinalis*)
Zones 2–9

Hummingbirds love the bright red of cardinal flower, a native species whose range extends from the Southwest to eastern Canada, as well as south through Mexico and Central America. This perennial is widely cultivated, not only for its hummingbird-attracting qualities but also for its sheer beauty, though it tends to need some hands-on care as well as foresight in placement and planting.

Cardinal flowers need both moisture and partial shade; they can die off quickly if neglected during dry, hot weather. Given proper care and placement, however, they are easy to grow, if short-lived. They are late-season bloomers, making them valuable for attracting hummingbirds from high summer into early autumn. In my own experience, they like early-morning sun and water, then significant shade from midday through afternoon, and more water in the evening during the hottest days. In predominantly cool regions, they benefit from more sunshine. Unless, perhaps,

ABOVE A stand of cardinal flower emblazons a woodland swale in Grundy Lake Provincial Park, Ontario.

OPPOSITE Cardinal flower is an especially valuable hummingbird magnet for cold-climate gardens.

you live in such a cool climate, don't plant them against south-facing reflective fences or walls, as they don't like the extra heat generated in such locations.

You can establish cardinal flowers from seeds or starts, and in both cases, use a bit of mulch around them to help moisture retention. Plant them about 10 inches apart. They also fare well in containers. Cardinal flowers can grow up to about 4 feet tall, and they can self-sow, like most native species, after the seeds develop (but note: hybrid cultivars may not self-sow successfully). If your cardinal flowers develop seed pods in the fall, you can carefully pluck a few off and scatter the tiny seeds over the ground, but the plants are likely to do so on their own. These brilliant red flowers can also be deadheaded, though the regrowth flower stalks are usually far less robust than the original; otherwise leave them be, so plants can self-sow. After the growing season, add just a thin layer of fine mulch underneath the lowermost leaves for winter protection.

CHUPAROSA (*Justicia californica*) Zones 9–11

This native beauty, indigenous to the deserts of southern California, Arizona, and northern Mexico, is very attractive to hummingbirds. Chuparosa (aka hummingbird bush, beloperone) is a summer-deciduous shrubby perennial that grows in sandy or gravelly washes of the desert floor. It has succulent leaves, most of which drop away by late spring. From spring to fall, the plant bears colorful, bearded tubular flowers, typically red, sometimes yellow, that are magnets for hummingbirds. If not for its inability to thrive beyond desert environments, chuparosa would be a staple for hummingbird gardens everywhere. In the Desert Southwest it has become a favorite for xeric gardening and is available through regional nurseries and gardening retailers.

Chuparosa makes an excellent border shrub or stand-alone plant, growing to about 6 feet tall and wide, spreading by rhizomes. With sufficient watering in a garden environment, the shrubs can grow quickly; they are excellent additions to gravel plots where they can share space with other xeric plants. During prolonged periods of subfreezing weather, the plant may die back to the ground, but the roots typically survive to sprout anew. Like most perennials of the Sonoran Desert, chuparosa

ABOVE Mexican honeysuckle (*Justicia spicigera*) attracts hummingbirds as readily as chuparosa.

OPPOSITE Starting in spring, chuparosa is bedecked in tubular flowers that are hummingbird magnets.

is drought tolerant and need only be watered about once per week (or even less frequently for well-established plants); once well established, it can survive weeks without water. Newly planted shrubs should be watered daily to help establish an extensive root system.

Justicia spicigera (Mexican honeysuckle, firecracker plant), equally hardy and attractive to hummingbirds, thrives in the same warm, dry climates. It is typically evergreen but will die back during harsh winters. Native from Mexico south through most of Central America, Mexican honeysuckle produces clusters of beautiful orange tubular flowers. This shrub enjoys a bit of shade and needs to be watered only sparingly, though regularly, once established. Tough and resilient, both species make great accent shrubs or natural-looking hedgerows.

COLUMBINES (*Aquilegia*)
Zones 3–9

Columbines are iconic, among the showiest of native wildflowers, evoking images of montane wilderness. Native columbines are found throughout North America, and virtually wherever native species flourish, hummingbirds feed from them. This continent is home to about two dozen different columbine species, several of which are widely known to outdoors enthusiasts.

ABOVE Columbines are early bloomers, providing nectar for hummingbirds before many other flowers appear.

OPPOSITE Western red columbine (*Aquilegia formosa*) is a hummingbird staple in the mountains of the American West.

The Colorado blue columbine (*Aquilegia coerulea*) is the gorgeous, large-flowered bluish violet columbine found at high elevations throughout the Rockies and associated ranges from the Southwest to northern Montana; the yellow-and-red western red columbine (*A. formosa*) ranges from California to Alaska and inland to the Rockies; red columbine (*A. canadensis*) is the only native columbine found in the eastern United States and eastern Canada.

In addition to native species, gardeners can choose from many different cultivars, though the native species are both hardy and extremely attractive to hummingbirds, perhaps more so than many cultivars. Columbines are easy to grow from seeds, but as with most perennials, starts are often preferable in order to establish plants in the garden more quickly. Moreover, seeds from cultivars have a maddening tendency to sprout lovely green leaves but not flowers. Columbines tend to be early bloomers—late spring or early summer, depending on climate and species—making them valuable assets for hummingbirds seeking nectar sources early in the nesting season.

CROCOSMIAS
Zones 6–10

The stunning plants of the genus *Crocosmia*, native to Africa, have long been the subject of intense hybridization—to the tune of some 400 cultivars. Crocosmias (aka montbretias) are easy to grow by planting commercially available corms or plant starts in spring. Cluster a few corms together, about 6 inches deep, and space each cluster by about a foot. They'll sprout as narrow, spiky leaves, and then shoot up to produce bundles of attractive swordlike leaves, typically 2 to 3 feet long, with rigid, gently curved flower stems. Blooms begin early to midsummer and can last into early autumn.

Crocosmias proliferate in slightly acidic, nutrient-rich soil and require regular watering. Once established, they need little more than water; in winter, at least in cooler climates, a little mulch over the beds can help protect the dormant plants. In hot climates, mulch your crocosmia beds to help retain moisture. They grow in climates lower than zone 6 but usually as potted plants that can be taken inside in winter. Some growers in cold climates dig up the corms and store them inside, then replant them in spring, after the frost season

ABOVE Crocosmias make excellent fenceline plants, their vivid red or orange flowers dangling from gracefully arcing stems.

OPPOSITE Exotically beautiful, crocosmias will draw hummingbirds to your garden from early summer into fall.

is past. The only real downside to these hummingbird magnets is that you must cut and discard the dead/dying leaves at the end of the season, which can be laborious in large crocosmia plots; you may also find them becoming invasive, so if your crocosmia beds begin to flourish too much, simply dig in and divide the plants at their corms/roots. Replant the separated sections elsewhere.

In my zone 8 garden, crocosmias did extremely well occupying a narrow raised bed against a tall fence. They had direct south exposure, but summer's intensity was mitigated by shade from a vined trellis about 6 feet in front of the crocosmias, which seemed to appreciate the midday relief from the sun. The hummingbirds flocked to the flowers. Moreover, I once stuck a crocosmia in the ground where the dogs tended to romp, and that plant also thrived, leading me to assume that these beauties are so hardy and durable that even I can't kill one.

CROSSVINE (*Bignonia capreolata*) Zones 5–9

This colorful native flowering vine will climb more than 50 feet, vertically or horizontally, given the chance and time. Its 2-inch-long hummingbird-magnet flowers are often two-toned—red or red-orange with yellow throats, or yellowish with orange throats. Some cultivars are bright orange. Crossvine tends to be an early bloomer, providing a rich food source to hummingbirds in May and June, and often continuing into late summer. Its indigenous range encompasses the South, lower Midwest, and parts of Texas and Oklahoma, but it is now introduced in the wild in a few other places; like many vines, it is potentially invasive.

Once established in well-drained soil, crossvine is a vigorous, durable perennial, and one that is easier to control than trumpet creeper. It likes full sun but will tolerate shade, and it's resistant to drought. Beyond its native range and given its invasive nature, crossvine is best planted in a container; happily, it does very well planted in a large vessel fitted with or adjacent to a trellis, lattice, or arbor. It can also climb sheer surfaces, making it a terrifically decorative adornment to cement, adobe, or brick walls, or even old wooden barns and sheds. Crossvine can even

ABOVE **A 10-year-old crossvine, fully encasing its arbor, provides countless nectar-rich "ports" for hummingbirds.**

OPPOSITE **Profusely blooming crossvine is a surefire winner for attracting hummingbirds.**

be trained as a groundcover creeper, but keep your eye on it!

Crossvine is semi-evergreen, tending to retain its attractive leaves in the South through winter but dropping them in colder climates come fall, after they turn reddish purple. In zones 5 and 6, prepare the plant for winter by covering it with mulch as it dies back to the ground. Crossvines are easy to establish from cuttings, plant starts, or even seeds, and once the vine begins growing, caring for it is just a matter of regular light watering, plenty of sunlight, and pruning to keep it under control. Those pruned cuttings can be used to start new plants. In good conditions, crossvine will flower profusely, making it a hummingbird favorite in your yard.

FAIRY DUSTERS (*Calliandra*) Zones 9–11

The genus *Calliandra* includes more than a dozen species in North America. Many are commonly known as stickpeas (they are in fact in the pea family), but the shrubs most commonly called fairy dusters (aka powder-puffs) are the pink-flowered *C. eriophylla* (false mesquite), native to the southwestern United States and northern Mexico, and the red-flowered *C. californica* (Baja fairy duster), native to Baja and northern Mexico. In the Southwest, both are readily available at nurseries, less so beyond that region.

In the Desert Southwest where it is native, fairy duster is easy to grow and maintain. The common name derives from the beautiful fluffy

ABOVE These beautiful perennial shrubs easily make a Top 10 list of hummingbird attractors for dry gardens in warm climates.

OPPOSITE Fairy dusters are exotic-looking native beauties of the Desert Southwest.

flowers, with their delicate-looking tufts of numerous stamens; hummingbirds love them. These plants are relatively carefree, requiring little water, and little or no pruning; they are drought tolerant and can also withstand cold winter weather. Farther north, fairy dusters work best when planted in full-sun, heat-reflective areas, such as up against south-facing walls and fences. They are capable of blooming nearly year-round in the warmest areas, but the prime blooming season is late winter through spring. Fairy dusters make great stand-alone shrubs, border shrubs, and even container plants. They grow up to about 5 feet tall and wide, sometimes more, with the red-flowered *Calliandra californica* generally larger at maturity than *C. eriophylla*. Several exotic *Calliandra* species have been introduced, but because some of them can become invasive, the indigenous species are much better options for gardeners seeking to attract hummingbirds.

FUCHSIAS
Zones 6–11

So far, 122 species and subspecies of the genus *Fuchsia* have been cataloged by scientists. They are native primarily to Central and South America, with a few species native to Mexico, Hispaniola, New Zealand, and Tahiti. One of the most popular species, *F. magellanica* (hummingbird fuchsia), is widely hybridized with other species and has many cultivars. Virtually all fuchsias are attractive to hummingbirds, and many species and hybrids have long, slender, tubular flowers, unlike the more robust and extravagant blooms of *F. magellanica* and its many cultivars. Hummingbirds frequent both types of blooms.

Fuchsias are especially popular as hanging-basket plants because the blossoms of many species cascade downward, creating a luxurious burst of color in any garden, on a patio, or just about anywhere else. They are so popular in the United States that the dawn of summer is signaled by virtually every retail outlet having anything whatsoever to do with gardening putting out displays of fuchsia baskets for sale. Many years ago, one such basket led to my discovery of what should have been obvious to me: fuchsias do just fine as in-ground plants. I had a three-year-old fuchsia basket that had cracked, and because of my inattentiveness, the plant had withered to a few paltry-looking stalks. I couldn't bring myself to

Fuchsias, in their near-limitless variety, make beautiful garden accents—and hummingbirds love them.

discard it, so I planted it in the ground in a shaded corner of the backyard and watered it regularly. It adapted readily, recovered, and soon proliferated, growing to a height of 3 feet and sending out multitudes of blooms. Thereafter I learned that in-ground fuchsias are commonplace. Mine flourished for a decade and probably continued to do so after I moved away from that house.

In general, fuchsias like shade and water. They need sunlight, but in my experience, they fare best with morning sun but ample shade during the afternoon. Healthy fuchsias will bloom all summer, making them a valuable staple for your resident hummingbirds. Fuchsias prefer damp, well-drained soil. They are perennials but need winter care to assure survival in cooler climates. In-ground fuchsias need a thick layer of insulating mulch placed over the plant base. Potted and hanging-basket fuchsias can be overwintered indoors: hose them off carefully to remove insect pests, then store them in a garage or basement, or other such space where you can maintain an air temperature of 45 to 55 degrees. This will encourage the plants to enter dormancy, which will assure their long-term health. Give the plant a little water every few weeks. Bring the plants out of storage and into a bright, warm location about 30 days prior to the normal last frost date for your location; clip the dormant stems back by about half their length and begin watering regularly. Once the last frost has passed, move the plants outside.

ABOVE Many species, such as this Bolivian fuchsia (*Fuchsia boliviana*), have narrow tubular flowers ideal for hummingbirds.

BELOW Upright-growing fuchsia varieties can grow into robust, densely flowered shrubs.

HONEYSUCKLES (*Lonicera*)
Zones 4–9

The genus *Lonicera*, which includes the beautiful native trumpet honeysuckle, comprises at least 180 species worldwide. Among them are several prevalent garden species from Europe and Asia, such as common honeysuckle (aka woodbine; *L. periclymenum*), Japanese honeysuckle (*L. japonica*), and orange honeysuckle (*L. ciliosa*).

Japanese honeysuckle, though attractive to hummingbirds, is problematic; it grows very rapidly, sending out runners that will root and grow anywhere, and is invasive throughout much of the East and parts of the West. Woodbine has become

ABOVE Non-native honeysuckles, such as this orange honeysuckle (*Lonicera ciliosa*), should be planted in containers from which their roots cannot escape.

OPPOSITE Honeysuckles welcome all comers—and hummers.

invasive in parts of Oregon and Washington. In both cases, the invaders outcompete native plants for light, space, water, and nutrients, and the vines will twine around anything growing in close proximity, eventually covering small trees and shrubs, which can lead to their collapse. Most vining honeysuckles are capable of similar invasive feats, so the exotic species named here and their cultivars need to be contained in pots and planters in which their roots have no egress to the ground. So handled, and pruned or manicured as needed, they form beautiful vines on trellises and other supports, and readily attract hummingbirds. Use large pots and planters capable of containing a free-standing trellis, or place pots or planters against large in-ground trellises, arches, or arbors. You can also place large conical tomato cages directly into the flower pot. Properly contained, exotic honeysuckles make colorful and hardy additions to a hummingbird garden.

HUMMINGBIRD MINTS (*Agastache*) Zones 4–10

Most of the nearly two dozen species in *Agastache*, a genus of herbaceous perennials, are native to North America; several are widespread, others have limited ranges. With a name like hummingbird mint (aka giant hyssop), it stands to reason these flowers are hummer favorites. They are such desirable garden flowers that numerous cultivars have been developed, from white to deep purple and everything in between, including orange and red hummingbird-magnet shades, such as 'Tango', 'Red Coral', and 'Orange Coral', among others.

Agastache species and cultivars run the gamut from cold-tolerant varieties to xeric varieties. Once established they are easy to care for and are equally at home in the ground or contained in pots. They make great edges, bookends, fence liners, and walkway borders; grown in planters, they add a bright highlight to patios and porches. Hummingbird mints tend to do best in well-drained, rather infertile soils; in cold climates, they appreciate a

ABOVE Hummingbird mints bloom profusely and come in a wide variety of colors, including mixed shades.

OPPOSITE Hummingbird mints, as their common name suggests, are prime attractants to hummers.

southern exposure. Plant them on slightly elevated mounds to aid in drainage, using gravelly mulches around the plant base. As winter approaches, leave the gone-to-seed stems intact, and then snip them away when new growth is established in spring.

Of the several native species adapted to desert-like conditions, *Agastache rupestris* (threadleaf giant hyssop, sunset hyssop, licorice mint), indigenous to high-elevation areas of Arizona, New Mexico, and northern Mexico, is a sure bet for attracting hummingbirds. Its beautiful coral-orange flowers emerge from lavender buds, explaining why it has become popular in xeric gardening. Conversely, *A. urticifolia* (horsemint, nettleleaf giant hyssop) ranges through the Rocky Mountain West and from California and the Pacific Northwest northward to northernmost British Columbia; this hardy species, its blooms a mix of white or pale pink and light or bluish purple, can grow 6 feet tall and more. The purplish-blooming anise hyssop (*A. foeniculum*), a native to the Upper Midwest and Great Plains, grows to at least 4 feet tall.

The genus offers many other species, and cultivars are ubiquitous. Overall, the hummingbird mints that fall into the red/pink spectrum seem to be better hummingbird attractors than the blues and violets, but hummers (and bees) will visit them all.

LANTANAS
Zones 8–11

Native to the tropical Americas, lantanas add bedazzling splashes of hummingbird-attracting color to any garden in temperate and warmer climates. In most of the numerous multihued cultivars, the colors change after pollination or as each bloom ages. Lantanas are easy to grow in-ground or in containers; the plants like sun and well-drained soil, making them highly versatile additions to a hummingbird plot. Under ideal conditions, lantanas can be invasive; they are therefore best contained in lined pots or boxes, where they usually flourish. Also note that lantanas are toxic to livestock and dogs, so avoid if you own pets.

These perennial shrubs can grow to about 6 feet tall, and, once established, require only modest watering and little or no fertilizing (which can retard blooming). In areas where the plants grow profusely, they withstand pruning well. To keep them under control, prune the shrubs during late winter or early spring, cutting the stems back by about two-thirds of their height. Some lantana cultivars are seedless or nearly so, and so they don't benefit from deadheading. If your plants begin to form seed heads, then deadheading the

ABOVE Lantana flowers, borne in compact clusters, range from yellow to purple and are frequently bicolored.

OPPOSITE Largeleaf lantana (*Lantana camara*), just one of the species behind the many cultivars.

flowers can encourage reflowering; cut the stem a few inches below the spent flower. Moreover, in areas where the first frost comes late (if at all), and your lantana flowers have all faded by midsummer, you can encourage large-scale regrowth by aggressive pruning.

In cooler climates, where winter cold can kill lantanas, they are best as container plants because you can bring them indoors (in an unheated space, such as a garage with windows) for the winter. They will go dormant and then reawaken and grow in the spring. Be sure the containers get at least a bit of sunlight through a window. You can also grow these flowers as potted perennials in cooler climates by bringing the containers indoors to a well-lit, heated space ahead of the onset of the cool weather lantanas don't enjoy. For indoor/outdoor varieties, look for dwarf or weeping cultivars. They need only an inch or two of water per week (overwatering can cause root rot); move them indoors when overnight temperatures dip into the mid-50s. In climate zones where overnight temperatures routinely hover in the 50s at night, potted lantanas kept outdoors during summer do best when placed against south-facing walls where they can bask in early-morning warmth.

PENSTEMONS
Zones 3–9

With some 250 species, *Penstemon* is the largest genus of flowering plants endemic to North America. Penstemons (aka beardtongues) come in a full suite of colors, from white to deep burgundy, with many different shades of red, pink, blue, and purple. These beautiful flowering perennials, so universally popular with gardeners, have been the subjects of intense hybridization, not only for showy colors but for increased heat tolerance and disease resistance.

Many naturally occurring *Penstemon* species are available from native plant nurseries. The red-flowered species evolved to be pollinated by hummingbirds; one of the best for attracting hummingbirds is the widely cultivated firecracker penstemon (*P. eatonii*), but other reddish species are equally alluring to the hummers. Like *P. eatonii*, most of the red-flowered species are native to the American Southwest and Mexico, among them Hartweg's beardtongue (*P. hartwegii*), beardlip penstemon (*P. barbatus*), and superb beardtongue (*P. superbus*). In addition to the native species and

ABOVE In the wild, myriad *Penstemon* species, such as this mountain pride (*P. newberryi*), attract hummingbirds as well as bees and other pollinators.

OPPOSITE Penstemons come in many forms and colors; most cultivars are attractive to hummingbirds.

their selections, there is another class of garden penstemons, among them 'Red Ace', 'Firebird', and 'Scarlet Queen'; these hybrids tend to be less hardy than the native species (at least in terms of heat and drought tolerance), but they are incredibly showy when in full bloom.

The red-flowered penstemons, our chief hummer attractors, do well in warm weather, but most are short-lived, lasting up to about four years; in an ideal garden habitat and with care, they can last a bit longer. They like heat and can tolerate direct sunlight as well as partial shade, and do well in pots, especially when placed in sun-facing locations. Penstemons prefer well-drained soils, making them excellent choices for slopes and raised beds. Deadhead the individual flowers as they die off, and carefully prune away each spent flower stalk near the ground to encourage new shoots to grow. In fall, after the first frost, cut the stalks back to 3 to 4 inches and cover the plant base with a few inches of mulch.

RED HOT POKERS (*Kniphofia*) Zones 6–10

Red hot pokers (aka torch lilies) are characterized by long, slender, strappy leaves and beautiful red to yellow tubular flowers bundled tightly together at the end of rigid stalks. These dense clusters of colorful blooms are instant hits with hummingbirds, and in the proper climate zones, the plants are hardy and often profusely successful: they can send up flowers stalks some 4 feet tall, and left unchecked, the basal cluster of leaves can grow to more than a foot in diameter.

Red hot pokers are readily available in a range of bright shades. The most common look is a bright red on the upper blooms, transitioning to yellow at the lower part of the flower cluster. Because of their height, red hot pokers make great vertical accents to a garden; they can also be used to create

Hummingbirds and bees alike will swarm to red hot pokers.

ABOVE A yellow red hot poker does not lose its charm for hummers.

RIGHT *Kniphofia* is an African genus of flamboyantly exotic-looking perennials.

hedgelike borders and dense linings along fencerows or walls. Plants seem to do best when slightly elevated to allow for excellent drainage; they can develop root rot if sited in depressions where they sit in water.

When well established, red hot pokers send up numerous flower stalks simultaneously, though in my experience they struggle to continue blooming throughout the summer. For the sake of aesthetics, and also perhaps to stimulate new flower shoots to grow, the unsightly spent flower stalks need to be removed—I used to clip them at the base but soon realized the plant doesn't care if I just yank them out. For winter, you can cut back stalks and leaves, and provide a little mulch covering for the basal cluster, or simply let them be until spring, and then, as the first new leaves begin to emerge, clip off or pull out all the foliage from the previous year. You can also divide the clusters and replant the partial rootstock to establish new plants.

SALVIAS
Zones 7–10

The genus *Salvia* includes hundreds of species, many of them bearing the common name sage; all are attractive to pollinators. Hummingbirds feed from many different *Salvia* species, and you could hardly go wrong choosing varieties other than red, but several of the red species are among the best and most resilient hummingbird-attracting garden plants. Common names vary, but ideal *Salvia* species for hummingbird gardens are *S. microphylla* (one of the parents of Hot Lips salvia), *S. blepharophylla* (eyelash salvia), *S. darcyi* (Vermilion Bluffs Mexican sage), and *S. greggii* (autumn sage), and their cultivars. Each of these species is bright red with blooms that are easily accessible to hummers.

Salvia belongs to the mint family; most species are hardy and pest resistant, and many are aromatic, particularly as you prune them. Moreover, the species mentioned here are xeric, which means they are well adapted to dry conditions, so they thrive in warm and even hot climates, and frequently bloom all summer long. But note that in colder climates, the red *Salvia* species can be difficult to establish as perennials.

These *Salvia* species are subshrubs, meaning that they have a woody base, which can grow quite thick in older plants. The plants spread, slowly, via underground stolons, putting up occasional suckers, and they are usually easy to grow. In the proper climate zones—generally 7 to 10, but some cultivars are hardier—they can grow rapidly and bloom profusely, forming colorful, rotund shrubs that hummingbirds love. They take water sparingly, and most of the red varieties can tolerate full sun, though they may like a bit of shade during the day, especially in hot locations. Once well established, red-flowered *Salvia* species are durable—you can water them, leave home for a few days, and return to find them perfectly happy. Hot Lips salvia, a hummingbird magnet, is a bicolored hybrid with red-and-white flowers; don't be alarmed if the blooms transition to all white, as

Hot Lips salvia, an all-time favorite of mine, is popular with gardeners and hummingbirds.

they often do, especially in late summer. They usually return to red/white sooner or later. And sometimes the season's first blooms are all red.

Purple sage (*Salvia dorrii*) is native to arid regions of the Far West.

Salvias are most easily established from healthy starts. Plant them in a hole lined with an organic compost, and water thoroughly until the young plants are well on their way. If you plant more than one, space them appropriately as they can grow up to about 4 feet tall and wide. They can shade out nearby flowers, so give salvias some room. They are easy to prune, but mine have always done better if I prune judiciously, a little at time over the course of the summer. You can deadhead them to encourage new blooms, but it's usually not necessary. If salvias have an Achilles' heel it is that the hot-climate varieties, such as these red woody species, don't like long, cold winters in cooler regions. To guard against the occasional hard frost in my zone 8 garden, I'll cover the base of the plants with a layer of mulch. In the spring, when they begin to green up, I cut back any obviously dead branches.

Many *Salvia* species are annuals, including some very colorful species that readily attract hummingbirds. They last only one summer, or part of the summer, but they can add wonderful color and texture to gardens. Moreover, annual salvias are good choices for containers and with proper care—which generally means not much more than watering and judicious pruning—can bloom profusely, much to the delight of hummingbirds. Also, dozens of perennial and annual *Salvia* species are native to the United States and Canada, and some of these are routinely available from specialty nurseries. Among my favorites is the widespread purple sage (*S. dorrii*) of the West, which, during years of robust rain in the high deserts, can paint entire hillsides in brilliant royal purple.

SPOTTED JEWELWEED
(*Impatiens capensis*)
Zones 2–11

Spotted jewelweed is beloved by hummingbirds for its beautifully complicated, nectar-rich orange blossoms, which hang from long, delicate stems. This widespread North American native goes by a variety of other common names: orange jewelweed, common jewelweed, orange balsam, and orange touch-me-not. This last moniker, though it sounds ominous, derives from the plant's strategy for dispersing its seeds: explosive dehiscence. Spotted jewelweed's seed pods become very sensitive, and even the lightest touch causes five valvelike coils to forcefully eject the seeds.

Spotted jewelweed grows in moist, even wet, locations, often as an understory plant along river bottoms, bordering sloughs and swales, ditch rows, and other such places. It likes shade, but if the soil retains enough water, it can grow in full sun. If you have a perpetually moist, shady to partially shaded area in the yard, this native might be a good choice. Spotted jewelweed can spread rapidly through its aggressive self-sowing, but it's easy enough to thin out if need be, and a large, dense patch of this summer bloomer is a no-brainer for hummingbirds,

ABOVE Native throughout much of North America, spotted jewelweed is a late-season bloomer, making it a valuable addition to hummingbird gardens.

OPPOSITE Spotted jewelweed is one of the few annuals on this list of hummingbird favorites.

not to mention an effective shield against weeds that might otherwise claim space. Spotted jewelweed can bloom well into autumn and does well in cool northern climates. The plants have delicate succulent stems, easily broken, and shallow roots; you can pull them from the ground with the lightest touch. So they do best in relatively sheltered settings and will readily sprout up through other vegetation. If they begin to spread too aggressively, simply pull them up before they form seed pods.

However, in some states west of the Rockies, spotted jewelweed is classified as a noxious weed, as is the likewise invasive policeman's helmet (*Impatiens glandulifera*). Because of the ease with which nonnative *Impatiens* species can self-sow, westerners should avoid introducing them as part of a hummingbird garden. The Pacific Northwest has three native *Impatiens* species; if you are lucky enough to find them on your property, protect them (they do attract hummingbirds), but consider removing any nonnatives because they not only form dense thickets that eliminate habitat for native plants, but they can also hybridize with spotted jewelweed and our other native impatiens. Several excellent identification guides, as well as state noxious weed details, are easy to find online.

TRUMPET CREEPER
(*Campsis radicans*)
Zones 4–9

Native to most of the eastern United States and naturalized well beyond its indigenous range, trumpet creeper (aka trumpet vine) supplies abundant nectar to hummingbirds via beautiful, 3-inch-long, trumpet-shaped orange flowers. Capable of climbing more than 30 feet in a single season, this tough, resilient perennial vine can be grown from the ground or from large pots and planters outfitted with sturdy trellises, or placed against a trellis, arbor, post, fence, or even a wall. It dies back in the winter to begin climbing anew the following spring, and where it is hardy, trumpet creeper needs little if any winter care other than heavy pruning. Older vines fill out with dense creepers, leaves, and flowers, making wonderful adornments to trellises, arbors, and fences. Trumpet creeper thrives in a variety of well-drained soil types, including disturbed sites, and prefers full sun for blooming. Cultivars, bred for color and growth rate, are readily available.

These aggressive vines are available as seedlings and cuttings, and are easy to establish from either. With young trumpet creeper vines, especially those in containers, prune away early foliage to encourage the plant to put more energy into

ABOVE Trumpet creeper, here framing a waterfall at the Chicago Botanic Garden, makes an excellent addition to any hummingbird garden within its native range.

OPPOSITE Trumpet creeper is ambrosia to hummingbirds.

establishing a deep root system, which will benefit the vine handsomely in the long term. Containers with drainage holes provide ready egress for aggressive trumpet creeper roots, which send up new sprouts in undesired places, so monitor the drainage holes, place the container on a saucer-style base, or place the container on a hard surface, such as a patio. Water the vines daily until they are established; mature plants generally need water only once or twice a week, more during hot, dry spells, and a little during the winter in southernmost latitudes.

Trumpet creeper tends toward proliferation and as such is considered an invasive weed, readily colonizing new areas and potentially overwhelming other species. Be mindful of where you plant it and when in doubt, use containers, especially in regions outside the plant's native range. Prune and thin the vines throughout the growing season, and come autumn, cut them back to within about 10 inches of the ground. Skin contact with trumpet creeper can cause redness and swelling, so wear gloves, long sleeves, and sunglasses when working with or near it; and keep pets away from it as well.

TRUMPET HONEYSUCKLE
(*Lonicera sempervirens*)
Zones 4–9

Native throughout much of the eastern half of the United States as well as parts of Texas, Oklahoma, and Kansas, the spectacular trumpet honeysuckle (aka coral honeysuckle) is a hummingbird favorite long familiar to Ruby-throated Hummers, which share its geographic range. This hardy vine is also widely introduced, especially to eastern Canada, and is a popular garden flower in the West, where western hummingbird species find it highly appealing. Trumpet honeysuckle does best in warmer climate zones but is hardy enough to withstand harsh winters. In the Deep South, it is evergreen. Named for its trumpet-shaped reddish flowers, which bloom in clusters, this honeysuckle does well on trellises, fences, and arbors. Vines can reach a length of 20 feet or so, though they usually stop short of that; they can also be used as a sprawling groundcover.

Trumpet honeysuckle is a valuable nectar source for hummingbirds because it blooms most profusely in spring and early summer. Moreover,

ABOVE Trumpet honeysuckle is native to the eastern United States.

OPPOSITE Trumpet honeysuckle offers hummers a valuable early-season source of nectar.

the late-season fruits (berries) are eaten by a host of songbirds, and the blooms are also pollinated by native bees. Trumpet honeysuckle can be cultivated from rootstock or seeds, but of course starts are the easiest way to get vines established in your garden. In northern climates with short growing seasons, plant them in warm locations: place trellises or arbors against south-facing walls or fences, and provide plenty of water. In more temperate climates, trumpet honeysuckle tolerates full sun to partial shade. Well-established plants need only moderate watering, and they do well with a heavy mulch covering at the base to protect from freezing in winter and for moisture retention in summer. When planting these vines, first build the arbor or trellis, then plant the vines 6 to 12 inches distant from the structure; use a soft, stretchy material to bind the stems to the structure.

'Major Wheeler' is a popular choice for attracting hummingbirds. Its gorgeous red flowers bloom profusely through most of the summer, and the vine fills in rapidly under prime conditions. This selection of the species is widely available, likes full sun, and once established can tolerate drought.

Trumpet honeysuckle is one of about 20 honeysuckles native to North America; other species are also valuable to hummingbirds and other pollinators, and some are available through nurseries specializing in native plants. In addition to the native honeysuckles, numerous exotic species have been introduced for gardening, but despite their beauty, they can become invasive.

FOUR

Hummingbirds of the United States

Identifying the hummingbirds that feed in your yard can be both fun and frustrating. Sometimes it's simple, but in some cases—and especially in some places—distinguishing between species can range from downright challenging to virtually impossible. Much depends on location, location, location.

Easterners have an advantage: the eastern half of the United States and most of southern Canada is the territory of a single species, the Ruby-throated Hummingbird. Along the margins of that vast territory—the western Great Plains and the Gulf Coast—other species occur, primarily in migration and sometimes as vagrants. But from the Canadian Maritimes west across the prairies, from New England and all along the Atlantic Seaboard, from the Great Lakes southward, if you see a hummingbird, it's almost assuredly a Ruby-throated.

Oh, to have it so easy in the West—in which region the case of the lookalike Allen's and Rufous Hummingbirds is especially onerous. The sizable summer/migration/wintering range of the Rufous Hummer

Costa's Hummingbird.

completely overlays the much smaller geographic range of the Allen's Hummingbird, which breeds (summers) along the Pacific coast from southern Oregon to San Diego County, California. The southern end of the Rufous Hummingbird's sizable breeding range barely overlaps the northern extent of the Allen's Hummer's breeding range. Moreover, the broad migration range of the Rufous includes virtually the entirety of the Allen's hummer's range. In and near the range overlap of these two nearly identical species, location is of little help in separating one from the other.

ABOVE + BELOW These photos show the difference in the tail feathers of the adult male Rufous and adult male Allen's Hummingbirds. The Rufous male (above) has distinct notches in the second tail feathers (the two feathers on each side of the center tail feathers). This notch is lacking in the adult male Allen's Hummingbird (below).

The adult, fully plumaged males of the two species are most easily distinguished from one another, but in this case, the phrase "most easily" intimates the impossibility of distinguishing between females and juveniles of the two species by color. An adult male Allen's and adult male Rufous are generally separable by the amount of green, as opposed to orange on their backs, but there is enough variation that some male Rufous Hummers simply cannot be separated from an Allen's on that basis alone. For several consecutive years, the dominant male Rufous in my yard sported a significant amount of green on his back: fully green backstrap and specks of green on the lower back. Living outside the range of Allen's Hummers, I could be quite certain this was not an Allen's based on location alone, not to mention its courtship display (which differs from that of the Allen's), but not so fortunate are the hummingbird enthusiasts who live along the coast of California and southern Oregon.

Beyond the green or lack thereof, another field mark provides certainty in identifying adult male Rufous and adult male Allen's Hummers, but it requires a lucky and detailed look at the bird's fanned tail, or better yet, a diagnostic photo of the tail. These hummers have a total of 10 tail feathers, or rectrices. Ornithologists number the rectrices one through five on each side (left and right) of the tail, with the tail tip feathers being ones (R1), and the outermost tail feathers being fives (R5). In the adult male Rufous Hummer, the R2 feathers each have a distinctive notch on the inner edge, near the tip. The adult male Allen's lack these notches, and their rectrices are slenderer, more daggerlike in shape. The tail feather notches are minor or missing on juvenile and female Rufous Hummers.

If all this already sounds mighty complicated, take solace in the fact that experienced hummingbird banders and expert hummingbird identifiers can do little better than the novice birdwatcher in distinguishing between Rufous and Allen's females and juveniles, not to mention a few other look-alikes, by observing them in the field. Only upon capturing the little birds and carefully examining various feather characteristics in hand can they differentiate them, a fact borne out by some of the best hummingbird identification guidebooks and online resources.

The Rufous versus Allen's conundrum is hardly the only hummingbird identification challenge in the United States. Females and juveniles of several species are difficult to identify, particularly during migration, when most species are routinely found well outside their breeding ranges. Female and juvenile Broad-tailed, Calliope, and Rufous Hummers are all quite similar.

Happily, enjoying hummingbirds does not hinge on being able to identify the sex and age of every hummer species at your feeders. The challenge of trying to do so is fun for some people, but insignificant to others, and neither viewpoint diminishes the joy of watching these fascinating little birds in action.

The following species-by-species account of the 15 hummingbirds that regularly breed in the United States is not intended to be a dissertation on successfully mastering the most difficult identification challenges. Instead, these descriptions provide ample detail to identify *most* male, female, and juvenile species. But note: juveniles are generally difficult to identify early in their lives; a bit later, in late summer and fall, when they've developed some plumage characteristics, they are easier to identify.

Significantly, only eight of these 15 species are both common and, to varying degrees, widespread. The other seven (the Arizona group, as I loosely call them) are generally confined to the Deep Southwest, especially the sky island mountain ranges of southern Arizona, southern Texas, and a few locations in southernmost New Mexico; these are the Blue-throated Mountain-gem and the Violet-crowned, Rivoli's, Broad-billed, White-eared, Lucifer, and Buff-bellied Hummingbirds. In addition to their Arizona strongholds, three of these species also breed in specific Texas locations, and one—the Buff-bellied Hummingbird—is not found in Arizona at all, but rather along the Gulf Coast of Texas and sometimes eastward. But the best chance of seeing four, five, or maybe even six on a single expedition is to visit the Arizona locations listed in the next chapter, especially the American Museum of Natural History Southwestern Research Station, Patagonia Lake State Park, Patagonia–Sonoita Creek Preserve, Paton Center for Hummingbirds, and Ramsey Canyon Preserve.

In addition to these 15 species with U.S. breeding ranges, several Mexican species show up regularly in the United States, usually in the Deep Southwest; reports of Caribbean species in Florida also occur. In the bird world, such rare visitors, wandering well outside their normal ranges, are called vagrants. Among them are such species as the Plain-capped Starthroat, which is seen almost annually in Arizona, along with the Berylline Hummingbird, a vagrant to Arizona with a few breeding records there, and the Mexican Violetear, which has been seen in several states and two Canadian provinces.

GLOSSARY OF HUMMINGBIRD TERMS

gorget. The patch of color on a hummingbird's throat, usually iridescent

pendulum flight. A courtship display in which the male hummingbird flies in a short pendulum-shaped arc in front of the female

postocular stripe. A stripe of white or other color trailing behind a bird's eye

primary projection. How far the primary flight feathers extend in comparison to the tail

scaling. Darker feathers against a lighter background that appear like columns of fish scales

sexual dimorphism. Differing plumage between males and females of the same species

shuttle flight. A courtship display in which the male hummingbirds performs agile side-to-side maneuvers in front of the female

spangles. The individual iridescent feathers that make up the hummingbird's gorget

subterminal band. A band of darker (usually) color dividing lighter tips and darker base section of tail feathers

THE BIG EIGHT

ALLEN'S HUMMINGBIRD (*Selasphorus sasin*)

Male Allen's Hummingbird.

Male Allen's Hummingbird.

Male Allen's Hummingbird.

Female Allen's Hummingbird.

NO DISRESPECT TO THE CALIFORNIA QUAIL, but I think the Allen's Hummingbird would have made a fine choice for the official state bird of California because, other than a tiny length of coastline in bordering southwestern Oregon, the Golden State comprises the entire breeding range of this energetic, tiny hummingbird. The two subspecies of Allen's Hummer are essentially identical; as its name suggests, subsp. *sedentarius* is nonmigratory from its range in southern California, while the more northerly and widespread type, subsp. *sasin*, winters in Mexico.

IDENTIFICATION:
Size: 3.25 to 3.5 inches. *Male:* warm rusty-orange flanks, breast band, face, neck, and tail; green back and crown, often with rusty-orange rump and sometimes with limited rusty-orange at the nape; brilliant metallic-orange gorget that can appear bright orangish scarlet to deep coppery-green; clean white breast band or patch below gorget, and clean white ventral area. *Female:* green above, white below, with light rufous-orange flanks and rump; throat variably speckled, sometimes forming a small partial orangish gorget; fanned orange tail feathers show black subterminal band and broad white tips on the three outer feathers. When visible, a narrow, easily overlooked orangish eyebrow stripe reliably separates female Allen's from paler female Broad-tailed Hummingbird. *Juvenile:* similar to female but duller overall, with highly varying degrees of speckling on throat, and often scaled green on the back.

STATUS AND RANGE:
Common summer resident along the Pacific coast from San Diego County, California, to Curry County, Oregon; especially widespread from southern California to the Bay Area, becoming somewhat less common northward and as distance from the coast increases; uncommon to rare inland California; fairly common winter resident in southern California. The first spring migrants arrive in the northern extent of their range in March, with the bulk arriving from mid-April through mid-May. Fall migration begins in midsummer; by early August, most birds in southwestern Oregon and adjacent northwestern California have departed.

HABITAT: Most common in coastal scrub and coastal residential areas, from urban to rural.

VOICE: In addition to the male's courtship vocalizations, Allen's Hummers make a variety of calls similar to those of the Rufous, including *chip* notes and aggressive rattling sounds used in confrontational situations.

BEHAVIORS: The male Allen's Hummingbird engages in a distinctive courtship display flight: the display begins with a series of pendulum flights in which he swings rapidly back and forth in a shallow arc, covering about 25 feet horizontally in each swing, uttering a buzzing note and ending each arc with a flare of the tail and a high-pitched rattle. These pendulum flights, five to 15 in a row, occur 6 to 8 feet or so above the female. Often the final arc transitions into a high-angled climb and then a high-speed plunge, culminating above the female. Allen's Hummers will pugnaciously defend territories, feeders, and flowers.

SIMILAR SPECIES:
Broad-tailed and Calliope may co-occur with Allen's in migration, but the breeding ranges do not overlap. Adult male *Selasphorus* hummers with all-orange backs are Rufous; within the geographical range of the Allen's, adult male *Selasphorus* with all-green backs are almost always Allen's. Many male Allen's specimens are only safely separable from nearly identical Rufous by the un-notched rectrix 2 (R2) tail feathers, and narrower outer tail feathers, especially R5. These characteristics are usually pronounced in adult males but very difficult to see in real time. Females and juveniles are generally not distinguishable in the field from juvenile female Rufous. Allen's and Rufous males have different display flights, which can distinguish them where their breeding ranges overlap.

ANNA'S HUMMINGBIRD (*Calypte anna*)

Male Anna's Hummingbird.

Male Anna's Hummingbird.

Male Anna's Hummingbird.

Female Anna's Hummingbird.

Female Anna's Hummingbird.

Juvenile male Anna's Hummingbird.

ONLY IN RECENT DECADES has the Anna's Hummingbird become the most abundant hummingbird of the Pacific states. Moreover, it is the only hummingbird that remains year-round in the Northwest. This colorful recent arrival has greatly expanded its breeding range, which was once limited to the mountains of southern and central California—a story told more fully in this book's opening chapter. So new to the Northwest is the Anna's Hummingbird that it has not had time to evolve a migration pattern southward for winter; so it simply stays put, adapting as need be to Northwest winters. The male Anna's Hummer is one of the singing species, proudly belting out his screechy courtship song at any time of year. As with other hummers possessing both a colorful gorget and a vibrant crown, the male Anna's, in good light, can beam like a colorful flashlight when the bird turns its head to the right angle to catch the light. It's enough to drive camera sensors batty.

IDENTIFICATION: *Size:* 3.5 to 4 inches. *Male:* brilliant fuchsia gorget and crown, both of which can take on a broad range of radiant reddish shades, depending on light conditions, including deep blood red, rose red, neon salmon red, hot pinkish, coppery, or even, in the absence of any good light, dark gray or blackish. The gorget tapers to points on the sides, but not the long mustachelike points of the closely related Costa's Hummingbird; Anna's male is the only U.S. hummingbird with a red crown. Dorsal surface is entirely dark green; green flanks and pale gray breast, usually with darker scaling; center tail feathers are green, outer tail feathers are dark gray/black. *Female/Juvenile:* green above; palc grayish flanks with green wash; pale grayish breast; younger juveniles grayish green on the back. Maturing male develops iridescent fuchsia-red spangles on the throat and crown; adult female has columns of dark spots on her white throat, often forming a small central gorget of red or coppery spangles. Broad, rounded tail feathers, with the two outermost having black subterminal bands and white tips. Female's white eyebrow stripe is variable.

STATUS AND RANGE: Common year-round resident from southern and western Arizona, southern and western Nevada, through most of California, through western Oregon and western Washington, to southern British Columbia; increasingly common year-round resident in towns along the eastern foot of the Cascades, probably due to feeders, and summer/fall sightings have routinely been documented as far eastward as Idaho, western Montana, Nevada, and Utah. Common winter resident throughout much of Arizona and eastward into New Mexico and Texas. Overall, the Anna's range is expanding, and sightings, especially late summer through winter, are increasing well outside the established range.

?
?
RANGE OF ANNA'S HUMMINGBIRD.
Breeding
Nonbreeding
Nonbreeding (scarce)
Year-round

HABITAT: Residential areas, parks, farmlands, foothills, valleys, coastal lowlands.

VOICE: Male Anna's song is a scratchy high chatter that can continue for long periods of time; sings at all times of year; calls from both sexes include simple high *tsip* notes and an abrupt *brrrt* chase call.

BEHAVIORS: Individual males often stake out one or more feeders and defend them, including occasional fairly aggressive encounters with the even-more-pugnacious Rufous Hummingbirds, which often end up dominating Anna's and other species. The male's courtship display is a long, high-speed dive, almost always toward the sun, in which he gains speed by flapping, which produces a trill, and then folds his wings against his body for maximum speed before checking his dive dramatically at the bottom with a loud squeak produced when he suddenly spreads his tail. He also performs shuttle flights in front of the female, with his shimmering gorget fully extended.

SIMILAR SPECIES: Female Anna's has a dingier breast/belly than female Black-chinned, Broad-tailed, Calliope, and Costa's; female Broad-tailed has pale rufous flanks and some rufous in the tail; much smaller female Calliope has slightly rufous flanks. Female Black-chinned has columns of very small spots on its white throat, but the throat is brighter and lacks a small iridescent gorget that is common in female Anna's; female Anna's generally doesn't wag her tail while feeding (female Black-chinned commonly does so).

BLACK-CHINNED HUMMINGBIRD (*Archilochus alexandri*)

Male Black-chinned Hummingbird.

Male Black-chinned Hummingbird.

Female Black-chinned Hummingbird.

Male Black-chinned Hummingbird with entire throat appearing black.

Female Black-chinned Hummingbird.

IN THE SAME GENUS AS THE RUBY-THROATED HUMMINGBIRD that breeds in the eastern half of the United States, the intriguing Black-chinned Hummingbird is a species at home in various habitats of the West. The male's mostly black throat gives this species its common name, but just the right turn of the head, in the proper light, startlingly reveals a narrow, rich blue-violet gorget bordering the black. Black-chinned Hummers are tail waggers—their tail is almost always twitching when they hover to feed, a behavior that can help distinguish females from similar species.

IDENTIFICATION: *Size*: 3.25 inches. *Both sexes:* long, very slightly decurved bill; at rest, the wingtips extend past tail. *Male:* metallic-green back; black chin/throat with blue-violet band at lower edge (not always visible); dull white underparts, darker flanks brushed with metallic green; dark tail feathers, central pair metallic green, some others with a glossy purple tinge. *Female/Juvenile:* subdued metallic-green upper body and drab gray underparts with light metallic-green flanks; outer three tail feathers tipped with white, central pair of tail feathers metallic green. Female's bill is slightly decurved, unlike that of the female Anna's, Costa's, and Ruby-throated Hummingbirds, but this characteristic can be extremely difficult to discern.

STATUS AND RANGE: Common breeding-season resident from northern Mexico and much of Texas northward through Arizona, most of New Mexico and Nevada, throughout Utah, western and southern Colorado, most of Idaho, eastern Oregon and eastern Washington, and into southcentral British Columbia; a disconnected breeding range extends along the Pacific slope from central Baja to central California and also extends along the northern Sierra Nevada. Migration range overlaps breeding range and also includes virtually the entire West from the Rockies to the Cascades and Sierra Nevada. Mostly winters in Mexico, with some birds wintering along the Gulf Coast of Texas.

HABITAT: Varies widely; montane; open juniper woodland; riparian zones in coniferous, oak, and mixed forest; hardwood-lined streams in sagebrush steppe and other desert habitats; oasislike locations in arid country; highly adaptable as long as there are food sources.

VOICE: High-pitched warbling song is seldom heard; chipping calls when chasing intruders; male's aerial courtship display produces a trill at the bottom of the dive.

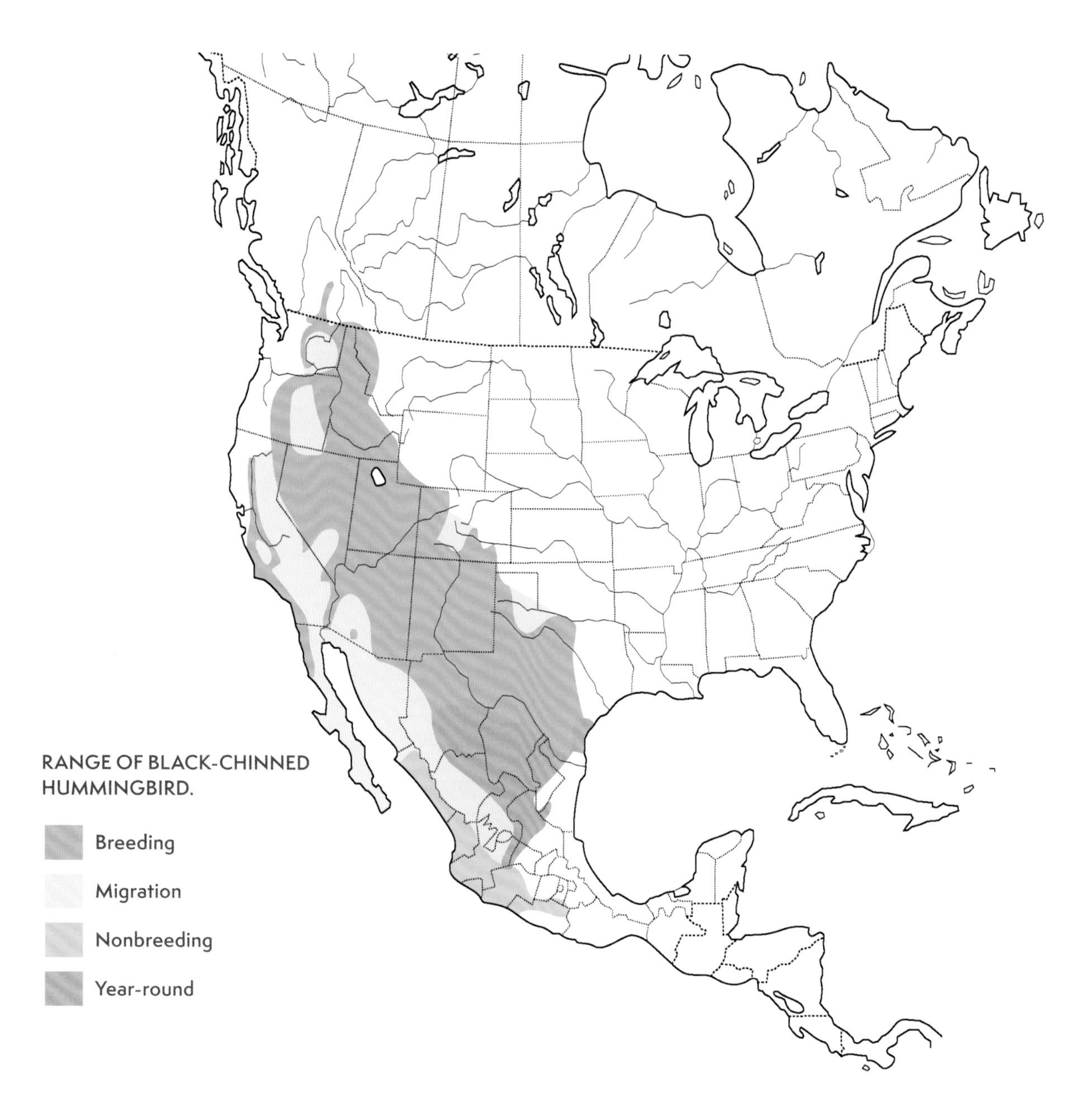

BEHAVIORS: Pumps and fans tail feathers often while feeding. Display flights are dives as high as 100 feet and also a flat figure-8 shuttle flight. Like some other species, Black-chinned Hummers often find a favorite exposed perch from which they can oversee their territory and will return to that perch repeatedly. Prior to southward migration, Black-chinned Hummingbirds often occur at high elevations, seeking flowers that bloom late in summer.

BROAD-TAILED HUMMINGBIRD (*Selasphorus platycercus*)

Male Broad-tailed Hummingbird.

Male Broad-tailed Hummingbird.

Female Broad-tailed Hummingbird.

Juvenile male Broad-tailed Hummingbird.

THE COMMON HUMMINGBIRD OF THE CENTRAL ROCKIES, the Broad-tailed Hummingbird reaches the northern extent of its breeding range in the mountains of southwestern Montana, eastern and central Idaho, and possibly far eastern or southeastern Oregon. Broad-tailed Hummers are the most common hummingbird species in Colorado and, during summer, are easy to find around feeders and areas of good habitat in many of the state's mountain towns. They breed and nest at high elevations, even above 10,000 feet. Thanks to the shape of its outer primary flight feathers, the male can produce a rapid, insect-like trill, which it uses during courtship display dives and during chase flights.

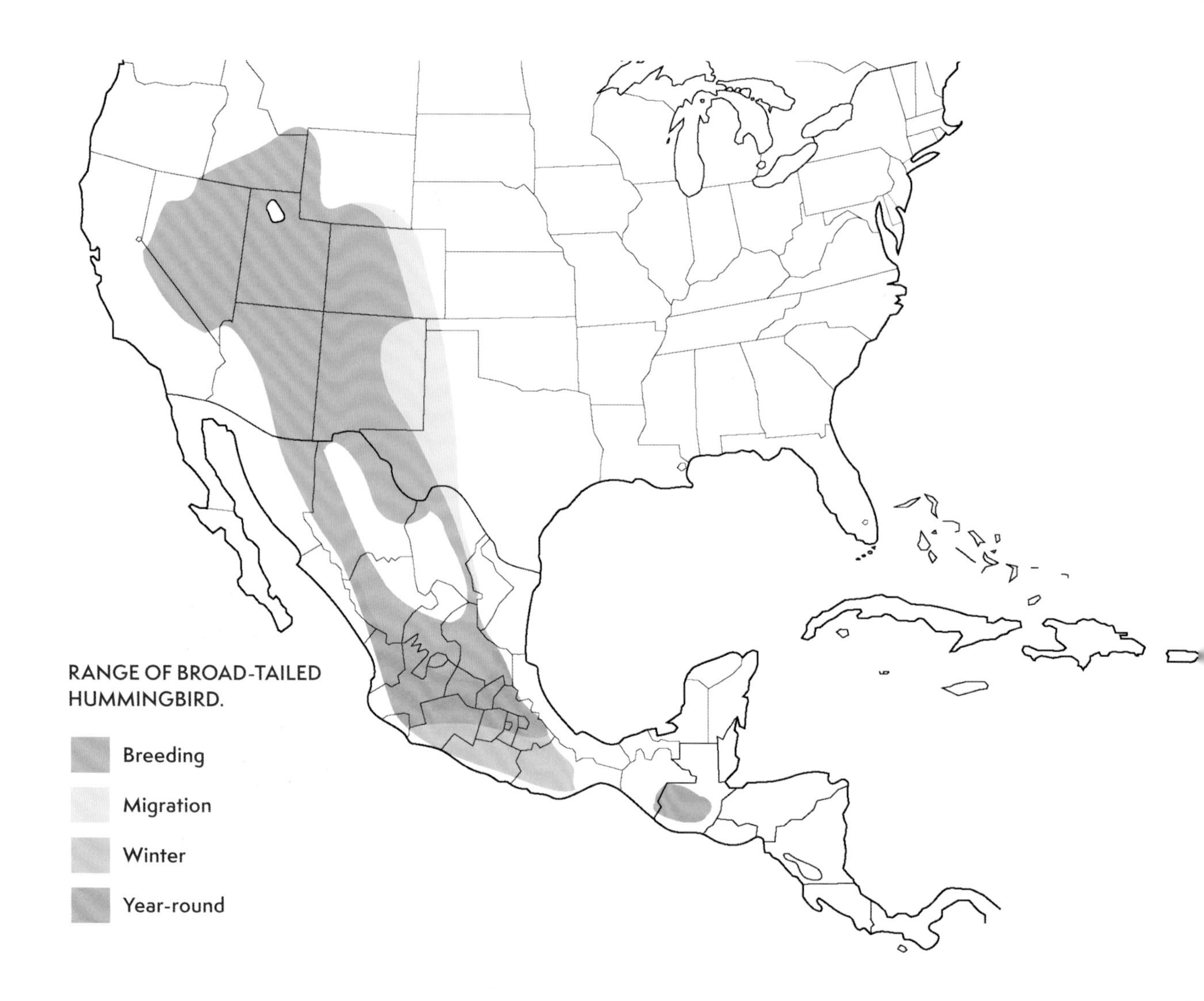

IDENTIFICATION: *Size:* 3.5 to 4 inches. *Male:* green above, including crown; brilliant rose to fuchsia gorget, which can appear substantially reddish or wine-colored; clean white below, except for greenish scaling on flanks. *Female/Juvenile:* difficult to distinguish from female Rufous Hummingbird, but larger and more robust, with less rufous in the base of the outer tail feathers, and paler with less rufous on the underparts; longer tail compared to wing length when perched (though this is difficult to judge in the field); when visible, the outer tail feathers are much broader than those of the female Rufous. Often perches less vertically than Rufous Hummingbird.

STATUS AND RANGE: Common to locally common throughout most of its range, but uncommon in the northern extent of its range across southern and central Idaho to southern and western Montana, except locally common in the Rockies along the Idaho/Wyoming border; rare in Oregon, northern Nevada, California. In migration, the range expands somewhat eastward, from eastern Colorado through central Texas; winters in Mexico.

HABITAT: Open meadows in mountain woodlands, including pine, pinion, juniper, pine-oak, aspen, aspen-conifer, mountain mahogany; brushy montane areas; riparian and nonriparian draws, ravines, canyons; common in populated areas in heart of range where feeders and preferred habitat are available.

VOICE: Typical calls are high, rusty squeaks; the male produces a loud, metallic wing trill in flight, which can help differentiate it from Black-chinned Hummingbird during high-speed flying maneuvers.

BEHAVIORS: In his courtship display, the male flies rapidly high into the air, where he then hovers, his wings producing a loud trill; he then dives speedily, headlong, down toward the female sitting below on a low perch, checking his dive at the last moment and whisking by her, and then once again climbs high in the air to start anew from the other direction. He also performs shuttle flights for the female. During migration, when Rufous Hummers use the same feeding areas, they often completely subjugate the less-aggressive Broad-tailed Hummers. Broad-tailed Hummers feed readily on sap from sapsucker wells—a key source of food, along with bugs, when flowers in the high country are just beginning to bloom in early summer.

SIMILAR SPECIES: Female and juvenile Calliope Hummers resemble female and juvenile Broad-tailed but are smaller and have very little rufous in the tail feathers. Female Calliope is tiny, with a very short tail that lacks any substantial rufous or orange, appears grayish overall, and has white tips on R4 and R5 (the two outermost tail feathers on each side); rufous wash over flanks is a very pale peach shade. Female Broad-tailed has less orange in the base of the outer tail feathers than female Rufous, making the tail appear predominantly greenish with extensive black then white bands, whereas the tail appears substantially orangish in the female Rufous; female Broad-tailed's tail usually extends noticeably beyond the folded wings, whereas the Rufous tail is about equal in length to the wingtips, but this characteristic can be very difficult to distinguish in the field; female Broad-tailed lacks a central cluster of spots on its otherwise lightly speckled throat.

CALLIOPE HUMMINGBIRD (*Selasphorus calliope*)

Male Calliope Hummingbird.

Female Calliope Hummingbird.

Female Calliope Hummingbird.

Male Calliope Hummingbird.

Juvenile male Calliope Hummingbird.

THE SMALLEST BIRD BREEDING NORTH OF MEXICO IN NORTH AMERICA, the dainty Calliope Hummingbird is a mountain specialist, breeding as high as timberline from the Sierra Nevada and Cascade crest to the Rockies. When subalpine meadows burst into colorful bloom, look for these unobtrusive little hummers hover-feeding at wildflowers that may not grow more than a few inches high. They are among the long-distance migrators, their annual path taking them as far north as the Canadian Rockies of central British Columbia from their wintering grounds on the Pacific side of central and southern Mexico. In fact, this circuitous journey, which can reach 5,000 miles for some individuals, makes the Calliope Hummer the world's smallest long-distance migratory bird.

IDENTIFICATION: *Size*: 3 inches. *Male:* light, bright green above; white below, with light green scaling on flanks; sparse, streaked magenta gorget with elongated spangles arranged in rows that leave gaps of white; short, stubby, all-dark tail can appear very slightly forked unless flared; wings extend to end of tail; juvenile male has only partial gorget. *Female/Juvenile:* green above, pale buff-tan below; white throat lightly flecked with green or bronze-green; short, stubby tail has black subterminal band and white tips on outer feathers.

STATUS AND RANGE: Locally common to uncommon (depending on location) summer resident at higher elevations in western mountains, including the Sierra Nevada and Cascades, and the Rocky Mountains from Colorado to interior central British Columbia and southwestern Alberta. In migration, Calliope Hummingbirds move northward primarily through the Pacific states; during their late-summer departure for the Mexican wintering range, they course southward through the Interior West, making their migration path an elongated oval.

HABITAT: Shrubby mountain meadows and riparian margins, especially open ponderosa forest, lodgepole stands, and mixed conifer forest, including 8- to 15-year-old regrowth from fire or logging; also montane deciduous-dominated stands (willow, alder, birch, aspen, maple) up to and above timberline; generally breeds above 3,000 feet but often found at lower elevations near breeding range upon arrival from wintering grounds; nests and visits feeders in mountain towns with appropriate habitat.

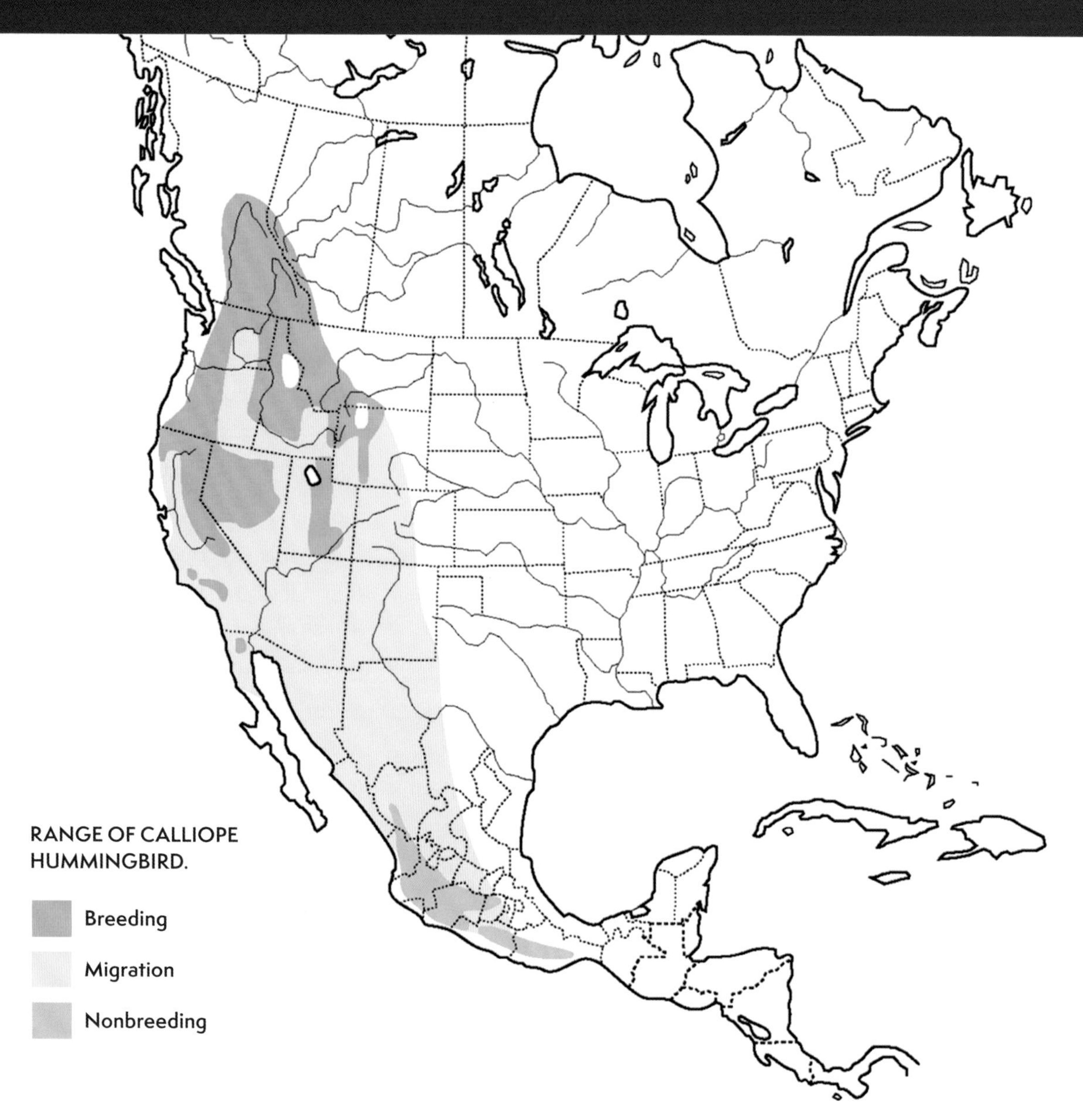

VOICE: Typical call is a very high, thin *tsip,* often doubled or in series; in courtship display dive, male utters a high spiraling *spizee* along with insectlike wing trills.

BEHAVIORS: Because many mountain flowers form low cushions, Calliope Hummingbirds often hover-feed near the ground; they also feed from sap wells drilled in trees by sapsuckers. The male's courtship display is a steep dive from aloft, along with hover displays progressively lower in elevation.

SIMILAR SPECIES: See Broad-tailed and Rufous for discussions of female plumage.

COSTA'S HUMMINGBIRD (*Calypte costae*)

Male Costa's Hummingbird.

Male Costa's Hummingbird.

Female Costa's Hummingbird.

Juvenile female Costa's Hummingbird.

Juvenile male Costa's Hummingbird.

THE STRIKING MALE COSTA'S HUMMINGBIRD is unmistakable, but the female is difficult to distinguish from females of other species, especially Anna's and Black-chinned Hummers. The adult male sports a brilliant royal purple crown and gorget, which envelops the face and base of the bill, and forms a long, pointed "mustache," making him one of the most photogenic of the always-photogenic hummingbirds of North America. Though Sonoran desert scrub is the Costa's natural habitat, vagrants appear annually in the San Francisco Bay area and almost annually from northern California to British Columbia. Virtually all reported vagrants are males; perhaps this is because males are more prone to wandering beyond the species' normal range or because the male Costa's Hummer stands out so readily from other species; vagrant females might occur as well, perhaps at a similar frequency, but would easily go undetected because they are so similar to female Anna's (and others): observers beyond the Costa's range presumably would not be scrutinizing female hummers to determine if any are Costa's.

IDENTIFICATION: *Size:* 3 to 3.25 inches. *Male:* iridescent purple gorget extends down to long tapers on both sides of the neck; iridescent purple crown; depending on the light, the gorget and crown can also appear deep bluish or black. Juvenile male is paler overall with minimally formed gorget. *Female/Juvenile:* green above, clean white below; often with flecks of purple on throat in the juvenile male and sometimes the adult female; short tail; wingtips extend to end of tail.

STATUS AND RANGE: Common during spring breeding season, with post-breeding dispersal and migration occurring in early summer; locally common year-round resident throughout most of breeding range; rare and local in northwestern part of range (e.g., Monterey County, California). Rare vagrant north into the Pacific Northwest.

HABITAT: Desert scrub, including urban and suburban areas of appropriate habitat.

VOICE: Abrupt high-pitched *tic* notes, sometimes in excited, rapid series; male's tail feathers vibrate at a critical speed during their display dive and emit a very high-pitched twirling whistle.

BEHAVIORS: Often pumps tail up and down while hovering at feeders; in breeding season, males often patrol their territory from one or more perches on the uppermost tips of shrubs or small trees. The male's courtship display includes a speedy U-shaped dive, during which the tail feathers vibrate to send out a high-pitched whistle, which rises in pitch as his speed increases; the males also flare their extensive gorgets in a bedazzling display of iridescence.

SIMILAR SPECIES: Female Anna's, Broad-tailed, and especially Black-chinned Hummers are difficult to distinguish from female Costa's, but all three are larger than the diminutive and compact-appearing Costa's. Female Costa's has a mostly clean white throat and light gray forecrown and cheek patches; it is generally less extensively green on the dorsal surface than the Anna's and Broad-tailed. Broad-tailed Hummer has a noticeably longer tail when seen perched. The Anna's female and juvenile generally have a small central spot of reddish gorget feathers that the female Costa's lacks, though she (and juveniles) sometimes have a few purple feathers on the throat.

RANGE OF COSTA'S HUMMINGBIRD.

RUBY-THROATED HUMMINGBIRD (*Archilochus colubris*)

Male Ruby-throated Hummingbird.

Male Ruby-throated Hummingbird showing distinctive forked tail.

Female Ruby-throated Hummingbird.

Female Ruby-throated Hummingbird.

Juvenile male Ruby-throated Hummingbird.

THE RUBY-THROATED HUMMINGBIRD IS THE ONLY COMMON HUMMINGBIRD in the eastern half of the continent, and the only breeding species throughout most of its range; unlike western hummingbirds, Ruby-throats tend not to swarm feeders and flowers except in certain places during migration. The Ruby-throated Hummer is also one of the most intensely studied species. Migration studies have confirmed that many Ruby-throats fly across the Gulf of Mexico in transit between the United States and wintering grounds in Mexico and Central America: staging in the Gulf Coast states in late summer, they add weight from extensive feeding and then undertake the 500-plus-mile nonstop journey over the Gulf, and then repeat this amazing flight on the return journey north in the spring, routinely arriving back on the Gulf Coast by early March and reaching the northernmost extent of their breeding range by mid-May. Ruby-throated Hummingbirds have the largest breeding range of any North American hummer, stretching some 2,000 miles from Florida to the Canadian Maritimes.

IDENTIFICATION: *Size:* 3.25 to 3.75 inches. *Male:* red gorget bordered by black chin and blackish face, then green crown; green back and rump, and green center tail feathers; white breast, light greenish gray flanks, white vent area; black tail is distinctly forked. *Female/Juvenile:* clean white below, including white throat and breast; developing juvenile male has fine columns of small streaks on the throat and sometimes a few spangles; back is green, and center tail feathers are green; outer tail feathers have broad black subterminal bands and white tips.

STATUS AND RANGE: Widespread and common, with a range extending from virtually the entire U.S. Gulf Coast northward to southern Canada, including the Maritime Provinces and westward to central Alberta, and from the Eastern Seaboard westward out onto the eastern edge of the Great Plains. The westernmost breeding range in the United States extends from eastern Texas and the east half of Oklahoma northward through eastern Kansas, and then through the easternmost parts of Nebraska and the Dakotas. In migration, Ruby-throated Hummers regularly occur west of the breeding range in the Great Plains. They primarily winter in Mexico and Central America; increasing numbers seem to be overwintering along the Gulf Coast.

HABITAT: Generally nests in open woodland, including gardens, ranging from deciduous forest to mixed conifer and conifer-dominated forests; common in residential areas, city parks, and other hubs of human activity with appropriate feeding and/or nesting habitat.

VOICE: Urgent, squeaky chattering calls are typical during chases in defense of territories and food sources; single-note *chips* and squeaky *chirps* are also common.

BEHAVIORS: In migration, especially fall migration, Ruby-throats gather by the many dozens at prime feeding locations throughout their range, providing an incredible spectacle for onlookers. Males will aggressively defend territories. They perform both dive displays and shuttle flights for females in the breeding season.

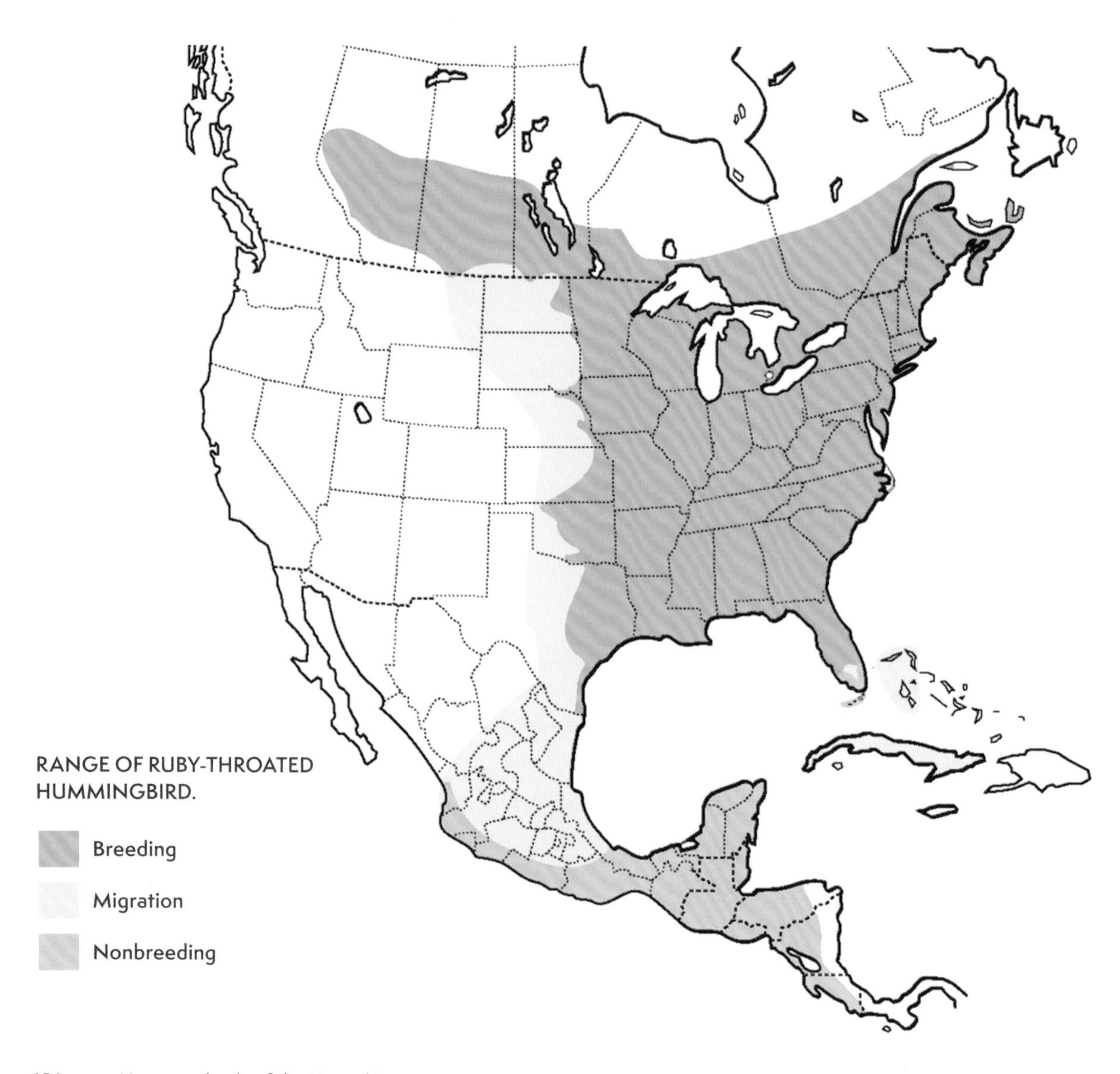

RANGE OF RUBY-THROATED HUMMINGBIRD.

RUFOUS HUMMINGBIRD (*Selasphorus rufus*)

Male Rufous Hummingbird.

Male Rufous Hummingbird.

Male Rufous Hummingbird.

Female Rufous Hummingbird.

Female Rufous Hummingbird.

Female Rufous Hummingbird.

THE UBIQUITOUS RUFOUS HUMMINGBIRD is the long-distance champion of the hummingbird world, breeding as far north as southeast Alaska and wintering in Mexico, and perhaps as far south as Panama, and along the Gulf Coast of the United States. During spring, Rufous Hummingbirds largely migrate northward west of the Rockies, and then return south through the same mountains. The migration paths of southern breeding populations, from northernmost California and western Oregon, are less well understood. Because of their flyway down the Rockies and southeasterly toward the Gulf of Mexico, Rufous Hummers tend to stray widely—much to the delight of birdwatchers in the eastern half of the country, who are annually treated to Rufous sightings, usually in late fall and winter.

In 2010, a one-year-old female Rufous Hummingbird caught and banded in Florida in January was recaptured in Alaska five months later; this incredible feat of continent-spanning flight was the longest hummingbird migration ever documented—a straight-line distance of 3,530 miles and likely substantially longer, as doubtless the little bird did not travel as the crow flies. Not only do these tiny dynamos return to the same wintering spots each year, but they tend also to gravitate right back to their favorite spots therein—their favorite flower patch, flowering bush, or feeder. Their northward migration begins early; the first birds reach the southern extent of their breeding range in southern Oregon in late February and the northern end of their breeding range in Alaska by late April. Spring all along the West Coast often confronts hummingbirds with chilly temperatures, frequently below freezing at night in British Columbia and Alaska, but Rufous Hummers are cold tolerant, routinely using torpor to survive frigid nights.

Rufous Hummers are equally at home on the coast and on inland mountain slopes. They can occur in the same breeding areas as Anna's, Calliope, Black-chinned, Broad-tailed, and Allen's Hummingbirds, but they don't particularly care to share feeders and will aggressively chase away, or attempt to chase away, interlopers of their own kind or any other species. For the time being they are the northernmost breeding hummingbird, but with the northward march of the Anna's Hummer into Alaska, they may soon need to share that title.

IDENTIFICATION: *Size:* 3.5 inches. *Male:* bright rusty-orange overall, often with flecks of green on upper back, sometimes with fully green patch across the upper or middle back; bright white breast band above buff-white belly; shimmering orange gorget in good light, but can also appear gold, greenish gold, coppery, or reddish, depending on light (or even blackish in bad light); green crown; pointed rusty tail feathers with dark tips, and the tail feathers adjacent to the two center tail feathers have a diagnostic notches on the inside tips (unlike male Allen's), but this feature is very difficult to observe in the field; juvenile male has rows of iridescent feathers on white throat and usually a central iridescent spot. *Female/Juvenile:* green above, dingy white below, with pale rust flanks; white throat with rows of small green spangles, often forming a small central gorget; outer tail feathers are rusty at the base, with black subterminal bands and white tips. Where their ranges overlap, female Rufous cannot be distinguished from female Allen's in the field and is difficult to distinguish from female Broad-tailed.

STATUS AND RANGE: Common and widespread spring/summer resident across breeding range from Oregon to southeast Alaska, and inland to the Northern Rockies northward from central Idaho; reports of breeding in northwestern California need further scrutiny. Generally common migrant; rare winter vagrant well outside normal range, especially in eastern states; locally common winter resident along the Gulf Coast from Texas to Florida.

HABITAT: Highly varied, from coastal scrub to alpine tundra, including open deciduous, coniferous, and mixed woodlands; brushy/shrubby fields, meadows, and prairies; riparian zones from desert to mountains; edges, clearings, meadows, brushy regrowth from clearcuts and fires; residential areas, city parks; widespread in migration.

VOICE: A variety of high-pitched notes and calls during the aggressive defense of territories, feeders, and flowers; calls include an excited, high, squeaky *zee-zee-chippity-chippity*, scratchy chattering *chip* notes, and high *zewee* notes. Males have a distinct insectlike wing trill in flight, whereas the female's flight produces a light hum.

BEHAVIORS: Frequents feeders, and both male and female aggressively defend feeders from fellow Rufous and other species; both sexes give warning *chip* notes, and if necessary will engage physically and noisily to drive away interlopers. The male's noisy courtship flight display is a steep dive from high above the female, ending with a very rapid J-hook at the bottom of the dive, with distinctive, buzzy, rapid *chew-chew-chew* calls; also zips back and forth in a rapid three-dimensional figure-8 shuttle flight in front of perched female.

SIMILAR SPECIES: Within the area of overlapping ranges, an adult male *Selasphorus* hummer with an all-green back is usually an Allen's, but most Allen's adult males have orange on the rump; a male with an all-orange back is a Rufous; but many Rufous Hummers have some green on the back. Sometimes the Rufous is only safely separable from nearly identical Allen's by the notched rectrix 2 (R2) tail feathers, and narrower outer tail feathers, especially R5 (difficult to observe in real time). These characteristics are pronounced in adult males, but not in females and juveniles, which are not distinguishable from Allen's. Allen's and Rufous males have different display flights, which—on the breeding grounds within the range of both species—can distinguish them. Rufous Hummers are more aggressive than Allen's, which can be dominated by Rufous at guarded feeders. Female Broad-tailed has less orange in the base of the outer tail feathers, making the tail appear predominantly greenish with extensive black then white bands, whereas the tail appears substantially orangish in the female Rufous; female Broad-tailed's tail usually extends noticeably beyond the folded wings, whereas the Rufous tail is about equal in length to the wingtips, but this characteristic can be very difficult to distinguish in the field; female Broad-tailed lacks a central cluster of spots on its otherwise lightly speckled throat. Female Calliope is tiny, with a very short tail that lacks any substantial rufous or orange, appears grayish overall, and has white tips on R4 and R5 (the two outermost tail feathers on each side); little to no rufous/orange on the body.

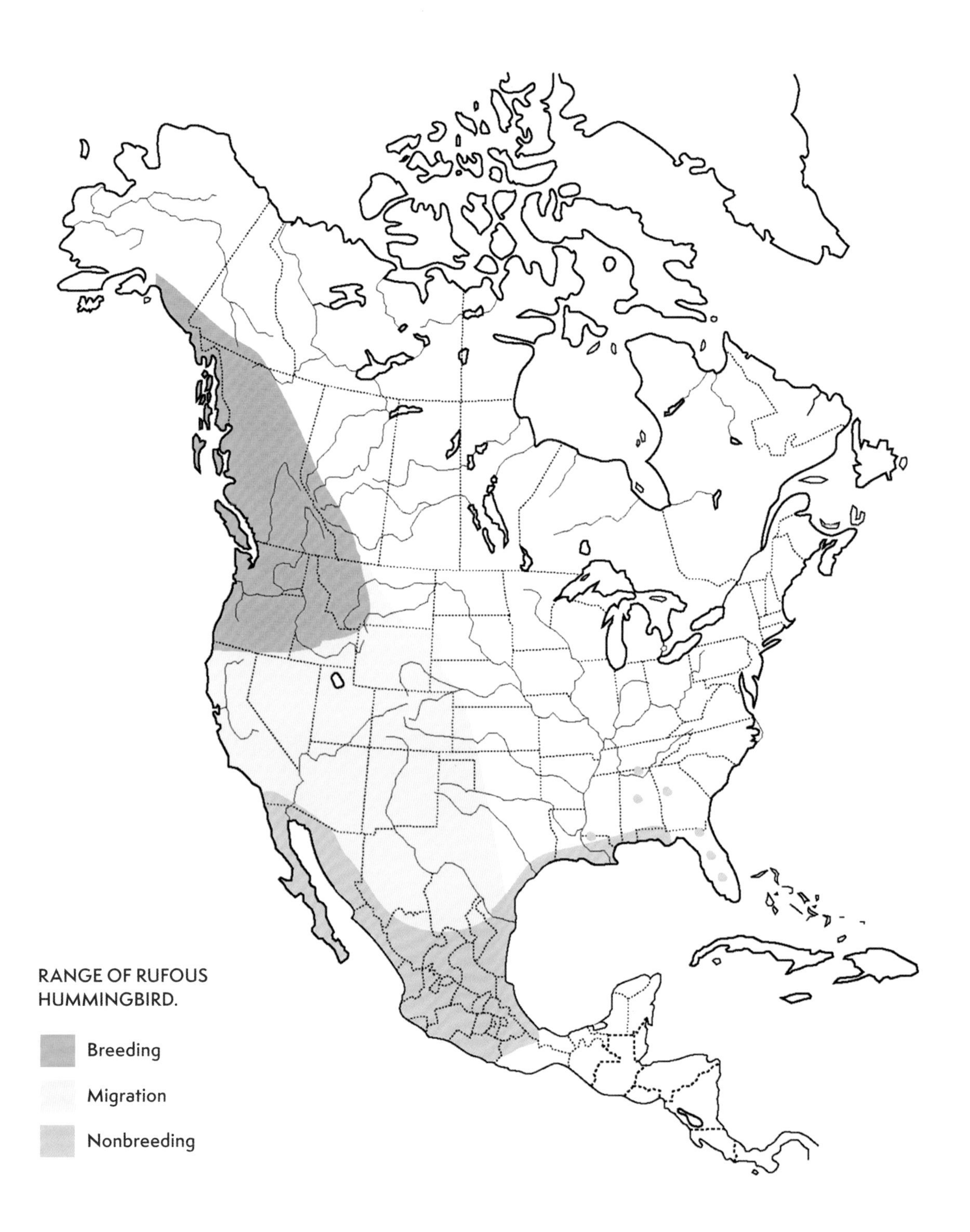

RANGE OF RUFOUS HUMMINGBIRD.

THE ARIZONA GROUP

EACH SPECIES IN THIS GROUP OF NORTH AMERICAN HUMMINGBIRDS has an extremely restricted range north of Mexico. Most are readily found only at well-known hummingbird hotspots in southeast Arizona—hence my moniker for them as a whole—but several species also occur in specific locations in adjacent New Mexico and in Texas. One, the Buff-bellied Hummingbird, is an uncommon but regular fall and winter visitor along the Gulf Coast, with limited breeding range in Texas, rather than the Desert Southwest. Beyond their very limited geographic ranges in the United States, these seven species are otherwise extremely rare vagrants north of Mexico.

BLUE-THROATED MOUNTAIN-GEM (*Lampornis clemenciae*)

Male Blue-throated Mountain-gem.

Male Blue-throated Mountain-gem.

Female Blue-throated Mountain-gem.

Female Blue-throated Mountain-gem.

THE COMPARATIVELY ROBUST BLUE-THROATED MOUNTAIN-GEM (aka Blue-throated Hummingbird) is the largest hummer species north of Mexico, weighing in at up to nearly 9 grams—or a bit more than a stack of three pennies. The Rivoli's Hummingbird is nearly as large; both species have comparatively low hovering wingbeat rates—about 23 beats per second. The Blue-throated Mountain-gem, whose U.S. breeding range is limited to the Deep Southwest, is one of the most habitat-specific hummers in the country, showing a strong preference for riparian zones flush with maple and/or sycamore. In such places, the male's distinctive *seep* calls are often the first sign of its presence, but Blue-throated Mountain-gems also use a wide array of other sounds, depending on the situation.

IDENTIFICATION: *Size:* about 5 inches. *Male:* bluish gorget in good light (can appear bright blue, pale blue, bluish violet, gray, or even black, depending on light); white postocular stripe begins above eye and trails back along the side of head; also a thin white stripe separating gorget from dark gray cheek patch; grayish crown; light gray breast/belly; grayish green back transitioning to long, wide, dark blackish blue tail with white-tipped outer tail feathers. *Female/Juvenile:* greenish above, transitioning into dark bluish black tail, with the outer tail feathers broadly tipped in white; white postocular stripe, pale gray throat, breast, and belly; juvenile males develop partial blue gorget.

STATUS AND RANGE: Locally uncommon at specific locations in southern Arizona (Chiricahua Mountains, Huachuca Mountains) and the Chisos Mountains of Big Bend National Park in Texas, with scattered breeding records elsewhere within this region, including New Mexico. During migration, Blue-throated Mountain-gems are found over a slightly wider geographic area, and a few birds overwinter in the region, generally at well-known Arizona feeder locations—Cave Creek Canyon, Madera Canyon, and Ramsey Canyon Preserve.

HABITAT: Breeding habitat in the United States is largely confined to high-elevation wooded riparian canyons featuring oak, sycamore, pine, and mixed species stands.

VOICE: Both the male and female have diverse vocalizations; common call of the male is a series of loud, somewhat piercing, evenly spaced *seep* notes; also a quieter buzzy chattering song. Males sing from favorite prominent perches, especially morning and evening.

BEHAVIORS: Females typically build their nests within 10 feet of the ground, occasionally attached to buildings, according to *The Texas Breeding Bird Atlas*. These large hummers tend to be aggressive and can dominate feeders, though they may ignore smaller species of hummers; breeding males are especially aggressive and engage in physical confrontations and chases with other males, and sometimes other species of birds.

SIMILAR SPECIES: The adult males are distinctive, but the females resemble the locally more common female Rivoli's. The female Blue-throated Mountain-gem is most easily distinguished from the female Rivoli's Hummer by the color of the central tail feathers: distinctly dark bluish black in the Blue-throated, green in the Rivoli's, with the lighter green central tail feathers contrasting somewhat with the blackish green outer tail feathers. Females of both species have white tips on the outer tail feathers, but they are usually much more prominently white on the Blue-throated.

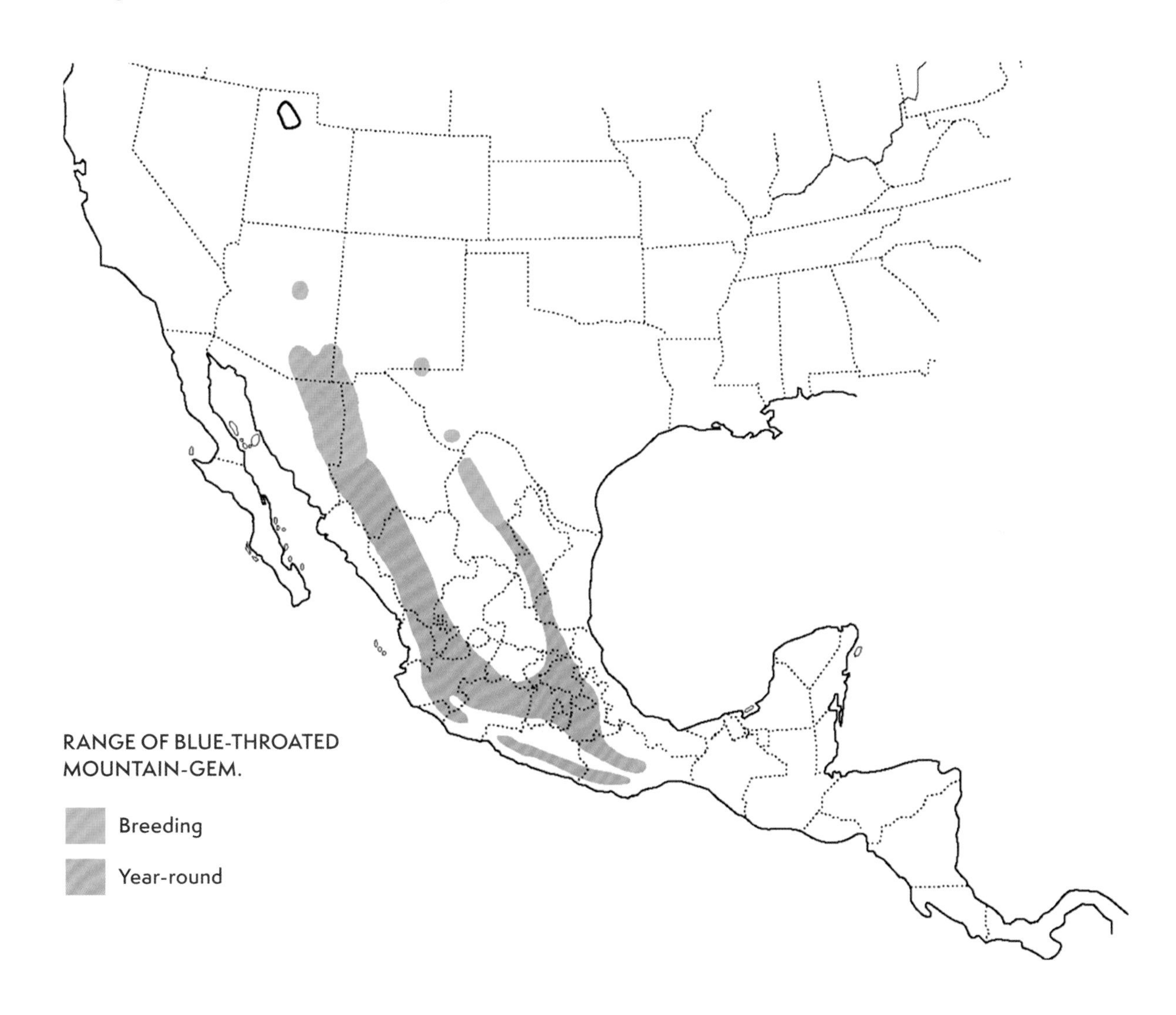

RANGE OF BLUE-THROATED MOUNTAIN-GEM.

BROAD-BILLED HUMMINGBIRD (*Cynanthus latirostris*)

Male Broad-billed Hummingbird.

Male Broad-billed Hummingbird.

Male Broad-billed Hummingbird.

Female Broad-billed Hummingbird.

Female Broad-billed Hummingbird.

Juvenile male Broad-billed Hummingbird.

COLORFULLY IRIDESCENT VIRTUALLY HEAD TO TOE, the kaleidoscopic little male Broad-billed Hummingbird is one of the most striking North American hummers. A member in good standing of the Arizona group, the Broad-billed Hummer is generally easy to find at the popular hummingbird hotspots in southern Arizona, where the appearance of a sparkling male in good light elicits awe and ahhhs from onlookers. In poor light, like the Rivoli's Hummingbird, he appears all dark, but he is substantially smaller than the Rivoli's and his bill is red rather than black.

IDENTIFICATION: *Size:* 3.5 to 4 inches. *Male:* unmistakable with his bluish gorget and breast, and bright red bill; dorsal surface is all green except for dark blue notched tail. *Female:* green above, with green center tail feathers, but outer tail feathers are bluish with white tips, most pronounced on the outermost tail feathers; green crown, dark gray mask over eyes with white eyeline above; whitish/pale gray below, including clean white throat; the bill is pink or red at the base, and the lower bill can be reddish for most of its length. *Juvenile:* similar to adult female; juvenile male has all-blue tail like adult male, except with narrow white tips on the outer feathers.

STATUS AND RANGE: Fairly common summer resident in southernmost Arizona, where birds often overwinter; rare in New Mexico and Texas.

HABITAT: In or near wooded riparian zones.

VOICE: Calls include a very rapid, high-pitched, chattery *chee-chit*, sometimes rapidly strung together machine-gun style, like a speedy, high rattle.

BEHAVIORS: Readily comes to feeders but tends to be nonaggressive and easily driven off by other species. Broad-billed Hummers are trapline foragers: they visit a series of favorite food sources rather than trying to dominate and defend a favorite nectar source. For courtship display, male engages in pendulum flights.

SIMILAR SPECIES: Female White-eared typically has distinctly black rather than gray face mask and larger, bright white eyeline; female White-eared also has scaling (spots) on white throat, and along the flanks, whereas female Broad-billed typically has a clean white throat with small spots only on the edges. Females of both species have red in the bill, but Broad-billed Hummer's bill is longer and thinner, and overall the bird appears slenderer.

RANGE OF BROAD-BILLED HUMMINGBIRD.

BUFF-BELLIED HUMMINGBIRD (*Amazilia yucatanensis*)

Buff-bellied Hummingbird.

Buff-bellied Hummingbird.

Buff-bellied Hummingbird.

THOUGH NOT NUMEROUS IN THE UNITED STATES, these bedazzling green-and-cinnamon hummers are increasingly regular winter visitors. Their overall range virtually encircles the Gulf of Mexico, from the Caribbean coast of the Yucatan Peninsula, up along the Mexican Gulf slope, and then through coastal Texas and Louisiana. They breed as far north as southernmost Texas but for some reason then expand their range in winter, rather than largely retreating southward like other species. The Buff-bellied Hummingbird is common from March through July and August in southern and coastal Texas; apparently individuals wander northward in late summer and fall after breeding.

IDENTIFICATION: *Size:* 4 inches. *Male:* green crown; green gorget and upper breast; gorget can appear grayish, depending on light; green back transitions to coppery-green, then cinnamon rump and tail; red bill with black tip. *Female:* a paler version of the male (Buff-bellied Hummingbirds exhibit minimal sexual dimorphism). *Juvenile:* like adult female, but with fewer spangles in the developing gorget.

STATUS AND RANGE: According to *The Texas Breeding Bird Atlas*, Buff-bellies breed in that state more often than their handful of confirmed nesting records would indicate: vagrants have been recorded along the eastern edge of the Edwards Plateau, along the coast to Louisiana, north to Austin and Travis counties, and as far west as Midland. In winter, individuals routinely occur in the Gulf States, becoming increasingly rare as the distance from Texas increases.

HABITAT: Open oak and mesquite woodlands, brushy areas, gardens and parks in towns and cities.

VOICE: Calls include a quick soft series of *tsee* notes, a more drawn-out *tseerip*, and a rapid combination, such as *tsee-tsee-tsee tseerip*; also, a mewing call, an intense high rattling call, and various *chip* calls.

BEHAVIORS: Breeding behavior is not well known. Regularly visits both flowers and feeders in several urban areas (Kingsville, Corpus Christi, Rockport) and national wildlife refuges (Aransas, Laguna Atascosa, Santa Ana) of coastal and south Texas in spring and summer. Provides birders along the Gulf Coast as far east as Florida an occasional welcome surprise when it appears at feeders during the fall and winter months.

SIMILAR SPECIES: The Berylline Hummingbird, rare in the United States, with only a few breeding records, has a gray, not buff, belly, and the upper surface of the bill is black; wings are bright cinnamon instead of dark. The U.S. ranges of the Buff-bellied (primarily Gulf Coast) and the Berylline (primarily Arizona) do not overlap.

LUCIFER HUMMINGBIRD (*Calothorax lucifer*)

Male Lucifer Hummingbird.

Male Lucifer Hummingbird.

Female Lucifer Hummingbird.

Female Lucifer Hummingbird.

IN THE UNITED STATES, THE LUCIFER HUMMINGBIRD is found in only a few places in the Deep Southwest; there it is a desert specialist, residing in dry, brushy canyons and arroyos, such as those incising the sky island mountains. In the Chisos Mountains of Texas, where their range and habitat overlap, Lucifers are frequently dominated by Black-chinned Hummers in defense of Havard's century plant (*Agave havardiana*).

IDENTIFICATION: *Size:* 3.5 to 4 inches. *Both sexes:* long, decurved bill. *Male:* gorget is brilliant purple-magenta, largely framed by white breast and white stripe trailing down the sides of the neck from the eyes; all-black tail is forked and often compressed to a point or double point rather than flared. Upper parts, including crown, are green; buff flanks. *Female/Juvenile:* green above, pale rust flanks, throat, and eyeline; black line from bill through eye gives a raccoonlike appearance to the face. Unlike the male's tail, the female's is multihued: the three outer tail feathers have white tips, broad black subterminal bands, and rusty-orange bases; the center tail feathers are green. Juvenile male is nearly identical to adult female but may show a developing gorget of purple spangles.

STATUS AND RANGE: Rare summer resident at key locations in western Texas, southern Arizona, and southwestern New Mexico—a region that forms the northernmost extent of its range. Almost all sightings reported on eBird and other online sites occur at the well-known hummingbird hotspots in these areas. Key sites are the Chisos Mountains of Big Bend National Park in Texas, where the bulk of the U.S. breeding population nests; also breeds in New Mexico and perhaps Arizona.

HABITAT: Arid mountainous terrain characterized by brushy arroyos, rocky slopes and canyons, and open scrubland slopes with a variety of important plants, including agaves and ocotillos. Readily visits feeders within its normal U.S. range.

VOICE: A variety of bright, loud, ringing *chip* notes, with numerous cadences.

BEHAVIORS: Unusual among hummingbirds, the male Lucifer frequently presents courtship displays to the female while she is building and sitting on her nest. The elaborate display begins with lengthy shuttle flights in front of her with gorget fully extended, tentaclelike feathers producing a crisp, light rattling; he then climbs high into the air to execute a high-speed dive.

SIMILAR SPECIES: Male Costa's has a purple crown and thin, short, straight bill; male Lucifer has a green crown and long, decurved bill.

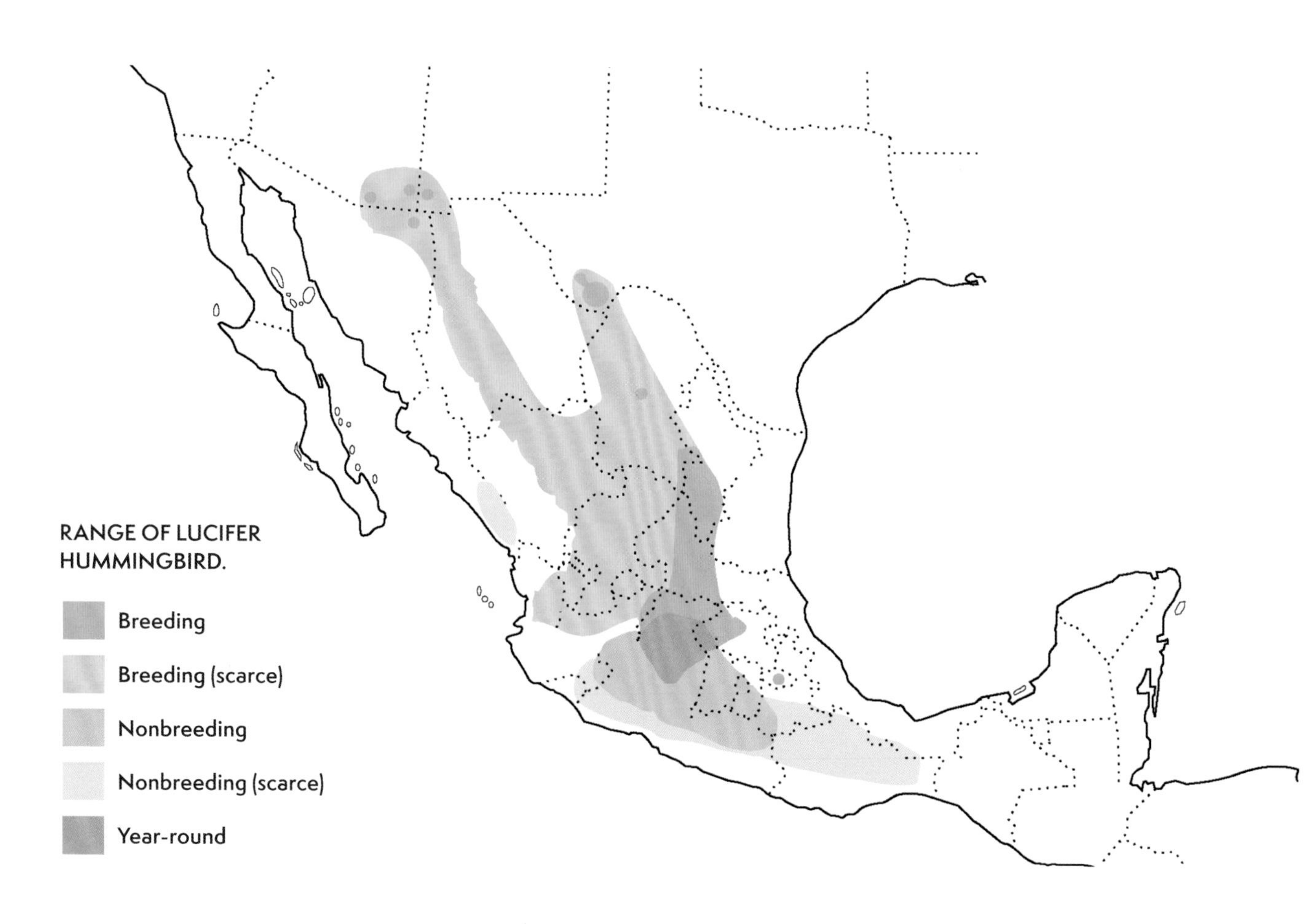

RIVOLI'S HUMMINGBIRD (*Eugenes fulgens*)

Male Rivoli's Hummingbird.

Male Rivoli's Hummingbird.

Female Rivoli's Hummingbird.

Female Rivoli's Hummingbird.

Juvenile male Rivoli's Hummingbird.

Juvenile male Rivoli's Hummingbird.

THE FORMER COMMON NAME FOR THIS SPECIES, Magnificent Hummingbird, seemed perfectly appropriate for a large, striking hummingbird that reaches the northernmost extent of its range in the American Southwest. But in 2017, the American Ornithological Society split the Magnificent Hummingbird into the Talamanca Hummingbird (*Eugenes spectabilis*) of Costa Rica and western Panama, and the Rivoli's Hummingbird, which ranges from northern Nicaragua to the southwesternmost United States.

IDENTIFICATION: *Size:* about 5 inches. *Both sexes:* fairly long, robust all-black bill. *Male:* black breast and emerald-green gorget separate it from all other U.S. hummers; also brilliant purplish crown separated from gorget by black face; in poor light, the entire head looks black and the entire bird is very dark, but at just the right turn of the head, the crown and gorget flash like a psychedelic strobe light. Body is predominantly green, with blackish breast; bronzy-green tail with no white tips. *Female:* greenish black tail with green center tail feathers and modest white tips on outer two tail feathers; green above, pale gray below with light greenish scaling on flanks, and columns of fine spots on light gray throat; white postocular spot/stripe. *Juvenile:* similar to adult female but somewhat grayer overall; juvenile males develop partial gorget in formative plumage.

STATUS AND RANGE: Uncommon overall in its U.S. range but locally common at well-known locations in Texas and Arizona; regular in the Davis Mountains of southwestern Texas; a regularly seen spring-through-fall resident at southern Arizona hummer hotspots; uncommon to rare farther north in the Santa Catalina Mountains near Tucson. Rare farther north and east (e.g., Mount Graham, Arizona; Signal Peak, New Mexico). Rare vagrant to Colorado, with scattered records elsewhere, outside normal range.

HABITAT: Generally a mountain species found in wooded regions, particularly, but not tied to, riparian zones; readily visits feeders.

VOICE: Male's song ranges from fast, rapidly repeated double- or triple-note *chip* calls, such as *chi-chip-cheer*, or *chi-chip-tee*, or *twi-chip twi-chip twi-chip*, to a series of single-note *chip* calls with changes in cadence. Calls include high-pitched kisslike *chip* sounds, sometimes stuttered together very rapidly.

BEHAVIORS: Studies summarized by the Cornell Lab of Ornithology suggest that in Arizona and other northern parts of their range, Rivoli's Hummingbirds tend toward nonaggression at feeders and flowers; rather than dominating a food source, they prefer to wander widely, using a trapline strategy of circulating among numerous nectar sources.

SIMILAR SPECIES: Female Blue-throated Mountain-gem has a sooty, dark bluish black tail with outermost tail feathers (R5s) almost half white and large white tips on R4s (tail color is diagnostic); also more defined and brighter white postocular stripe.

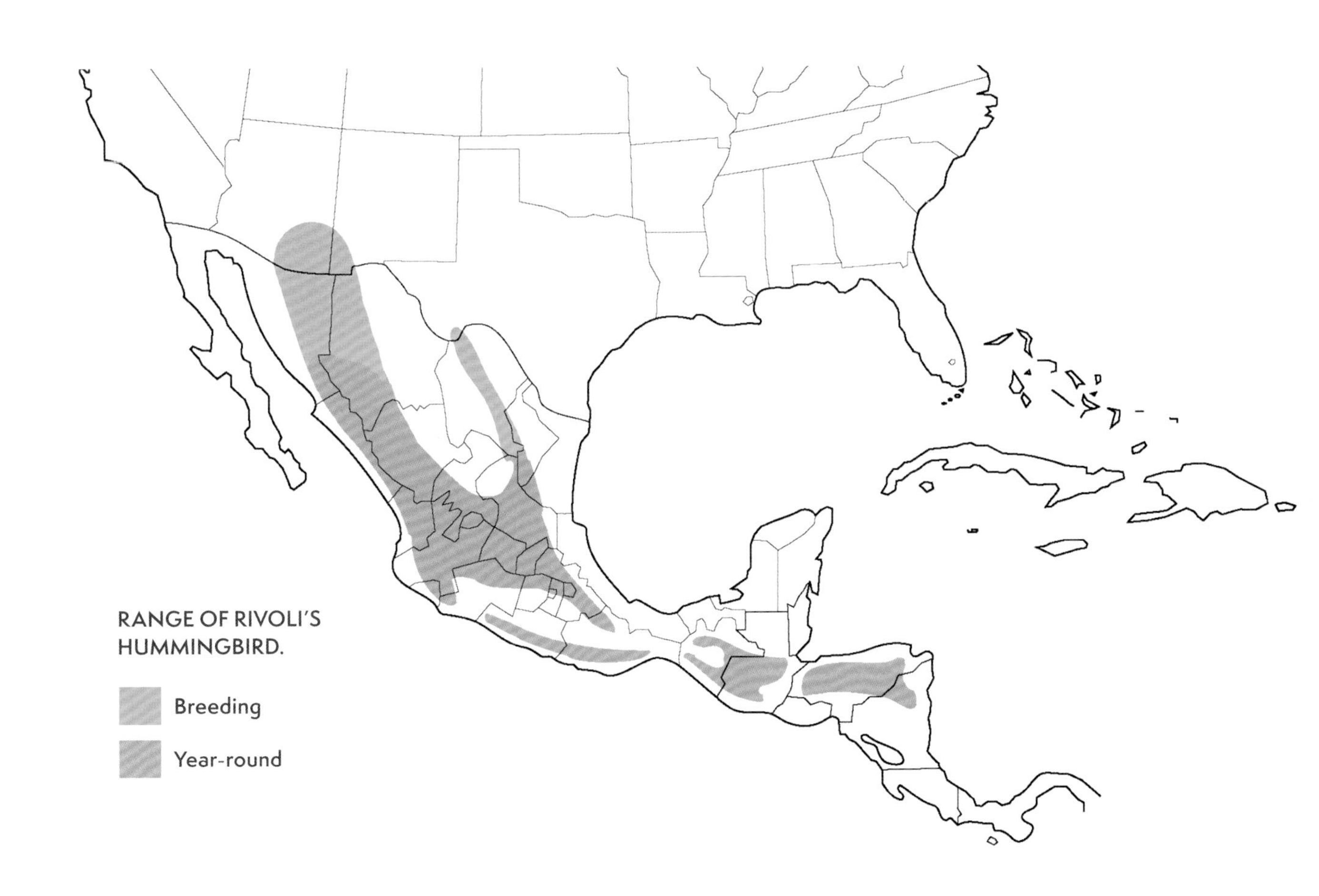

VIOLET-CROWNED HUMMINGBIRD (*Amazilia violiceps*)

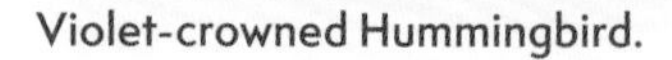

Violet-crowned Hummingbird.

Violet-crowned Hummingbird.

Violet-crowned Hummingbird.

UNLIKE THE OTHER SPECIES THAT BREED IN THE UNITED STATES, the enchanting Violet-crowned Hummingbird is sexually monomorphic—the male and female are identical. There is no way to differentiate between them in the field. They are also easy to separate from all other hummers in North America because of their bright white underparts and black-tipped red bill. The Violet-crowned Hummer is primarily a Mexican species, but its breeding range extends to southeastern Arizona and adjacent New Mexico, and the prevalence of feeders maintained by humans in well-known hummer locations within this region have increasingly encouraged this species to overwinter. Moreover, in the 20th century, the Violet-crowned Hummer expanded its range northward into Arizona and New Mexico—only a few sightings were recorded prior to the 1960s.

IDENTIFICATION: *Size:* about 4 inches. *Both sexes:* snowy white underparts; no other U.S. hummingbird shares the starkly white unmarked belly, throat, and flanks, which strongly contrast with the bronze-greenish back (which can appear grayish under some light conditions). The crown, nape, and face are purplish, sometimes appearing violet-blue or even more blue than purple, and varying in intensity among individuals. A bright red bill with a black tip adds a splash of color to these striking birds. *Juvenile:* paler, often appearing plain gray above and white below.

STATUS AND RANGE: Uncommon year-round resident and generally uncommon spring and summer resident in southeastern corner of Arizona, rare to locally uncommon in southwestern corner of New Mexico; but locally common and reliable at specific locations, primarily within its prime range in Arizona; rare to accidental elsewhere. Arizona/New Mexico populations are generally migratory, spending winters in Mexico and returning to the United States starting in March, then departing south by September; a few individuals routinely remain year-round at key sites, especially in and near Patagonia, Arizona. The bulk of the species' range is in Mexico, throughout most of which the Violet-crowned Hummer is nonmigratory.

HABITAT: Generally riparian. Nests in stands of Arizona sycamore (*Platanus wrightii*) and perhaps a few other species of deciduous trees; according to the Cornell Lab of Ornithology, almost all nests in Arizona and New Mexico are found some 20 to 40 feet above ground, in the outer branches of Arizona sycamores.

VOICE: Song is a loud, ringing series, *cheer-cheer-cheer*; calls, especially at dawn, include a kissing *chip-chip-chip*.

BEHAVIORS: Violet-crowned Hummers will actively chase away other species, especially in the absence of multiple nectar sources, and can dominate smaller species; often selects a favorite perch to use routinely. Feeds on a wide variety of flowers and readily comes to feeders.

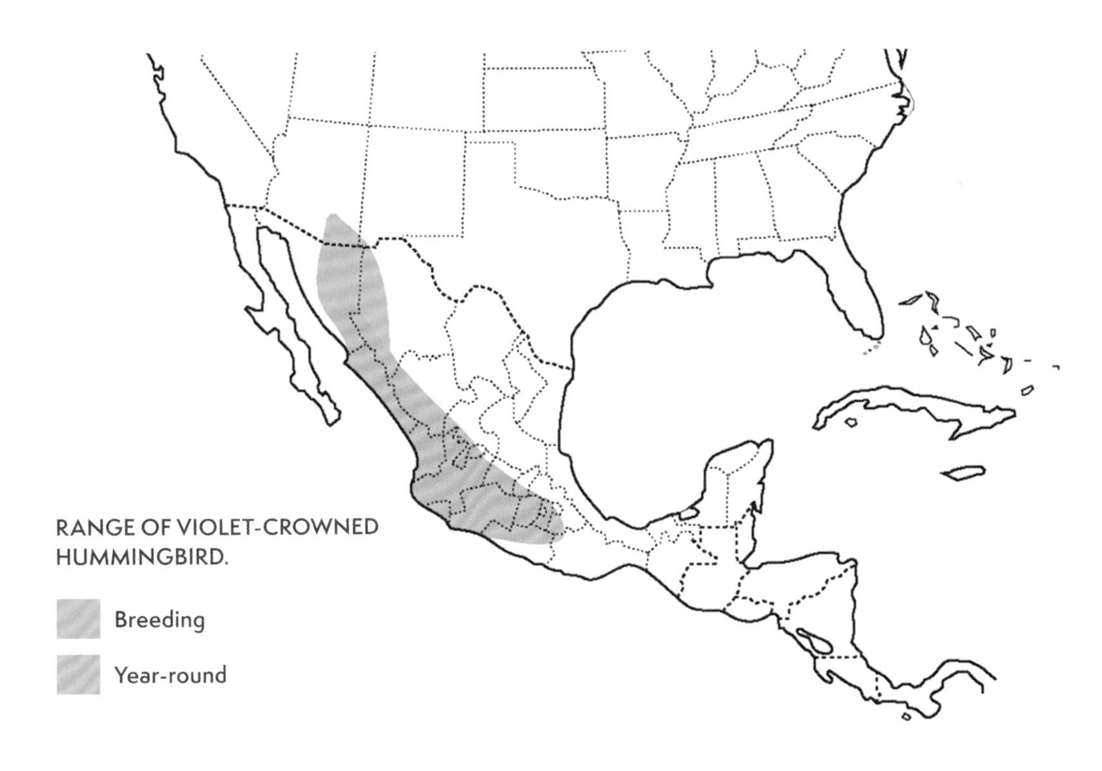

RANGE OF VIOLET-CROWNED HUMMINGBIRD.

WHITE-EARED HUMMINGBIRD (*Hylocharis leucotis*)

Male White-eared Hummingbird.

Female White-eared Hummingbird.

Female White-eared Hummingbird.

THE WHITE-EARED HUMMER IS A COMMON SPECIES in most of its Mexican and Central American range, but in the United States, it occurs regularly only in southern Arizona. When in 1996 the Arizona Bird Committee reported a confirmed nest, they called the White-eared a "rare but regular" summer visitor to southeastern canyons, particularly Ramsey, Carr, Cave Creek, and Madera, restricted to oak or pine zones at elevations of at least 5,000 feet; this summary appears to remain accurate regarding the relative abundance of the species in Arizona. The best hope for enthusiasts to see a White-eared Hummer is to monitor eBird reports from the key sites in Arizona and be ready to hop on a plane, so to speak: when a White-eared Hummingbird shows up at these locations, he or she tends to hang around for a while.

IDENTIFICATION: *Size:* 3.5 to 4 inches. *Male:* bluish purple face, startling in good light, encompassing the entire base of the black-tipped red bill, and then bordering a black crown and black eye patches, with showy bright white postocular stripes trailing down nearly to the shoulders; the striking purplish face borders a brilliant blue-green gorget and breast; the flanks are extensively scaled green, leaving only a narrow white belly patch; dorsal surface all green, with green center tail feathers and blackish outer tail feathers. Frequently, depending on light and angle, the male's entire head appears black (except for the white stripes). *Female/ Juvenile:* similar overall to the male but lacks the gorget; white throat with columns of dark (greenish) spangles; brownish crown; tail is bluish with green central tail feathers; outermost tail feathers have conspicuous white tips, less so on the adjoining feathers.

STATUS AND RANGE: Rare but almost annual spring/ summer visitor to southern Arizona, seen at the well-known mountain-canyon hummingbird feeder locations; considered a rare breeding species in the Huachuca Mountains near Sierra Vista; very rare visitor to southern New Mexico and the Davis Mountains of southwestern Texas.

HABITAT: Sightings in the United States occur at the Arizona hummer hotspots, such as Ramsey Canyon Preserve and Madera Canyon; in general it prefers wooded uplands with clearings.

VOICE: Male's song is a doubled whistling *cheer-ip cheer-ip cheer-ip* and sometimes a lengthy series of plaintive, clear, *chip* notes; the song can go on for quite some time. Calls include metallic *chink* notes.

BEHAVIORS: Combative in defense of food sources and territories, though subordinate to larger hummingbird species. Intriguingly, in their Mexican and Central American range, White-eared males gather at lekking sites, meaning several (up to seven have been documented) males gather to sing for females in a confined area.

SIMILAR SPECIES: See Broad-billed Hummingbird for comparison of females.

FIVE

Hummingbirds on the Road

Viewing hummingbirds up close is a mesmerizing and fascinating experience, an activity that appeals to young and old alike. Hummingbirds are so popular that many state and federal wildlife refuges, state parks, and other public facilities routinely maintain hummingbird feeders, much to the delight of visitors. Even private businesses like bed and breakfasts and restaurants maintain feeders as an amenity, to attract hummingbirds and entertain their guests.

The hummingbird viewing experience at publicly accessible venues varies considerably with location. Hummingbirds abound in the West, being so common in many areas that finding them often requires little more than keeping your eyes and ears open in the outdoors. This is especially true in the hummingbird hotspots of southern Arizona, and parts of New Mexico and Texas. At these places, visitors frequently see multiple species in prodigious numbers. In fact, visitors are essentially guaranteed to see many individual hummingbirds at the well-known hummingbird hotspots in the Desert Southwest during the spring and summer. Because they are so numerous, the hummingbirds of the

Multiple feeders at the Mill Restaurant in Arizona's Prescott National Forest.

Southwest draw tourists from all over the world. Nowhere in the United States is hummingbird tourism more evident than at the Arizona hotspots, such as the Paton Center for Hummingbirds in Patagonia.

But throughout the West, habitats from sea level to above tree level are often abuzz with hummingbirds. In the Mountain West and all along the Pacific coast, feeder locations can also be very busy, attracting many dozens of hummingbirds daily, especially at peak feeding times; even along Alaska's Prince William Sound, many if not most lodges put out hummingbird feeders for their guests to enjoy during the summer. But unlike many other states, the Pacific Northwest, California, and the Rocky Mountain states have comparatively few public locations where feeders are maintained, largely because (depending on location), Allen's, Rufous, Anna's, Black-chinned, Calliope, and Broad-tailed Hummingbirds are often common and evident, fairly easily encountered in areas with lots of flowering plants.

In the eastern United States (including the Great Plains states and upper Midwest), where just a single species, the Ruby-throated Hummingbird, holds sway, activity is usually far less frenetic at feeders. Sometimes just one or a few birds use feeders at visitor centers, park headquarters, and other public locations, especially during the summer breeding season, or as brief stopovers during migration. However, at locations that have maintained robust feeding stations for many years, migration can deliver swarms of Ruby-throated Hummers; and in the Deep South, along the Gulf Coast from Louisiana to Florida, fall and winter often bring surprise visits from a host of western species, along with the Buff-bellied Hummingbird, so many locations and individuals maintain hummingbird feeders year-round.

ABOVE Throughout the range of the Ruby-throated Hummingbird, it is especially important to call ahead to find out if the birds are using the feeders at locations you might want to visit.

BELOW It's worth considering—especially if you are introducing children to the joys of hummingbirds—that a single regular visitor to a feeder at a refuge or preserve is enough to plant the seeds of a lifelong fascination.

Wherever you go to see hummingbirds, time of year is paramount. In general, migratory species in the United States withdraw from northern latitudes between late July and early September and are by and large gone until the following spring, when, depending on species and location, they arrive between late February and June. Because summer/fall migration is southward and spring migration is northward, southern locales have hummingbirds both earlier in spring and later in summer/fall than more northerly places. In some places, certain species are year-round residents. You can now count on the Anna's Hummingbird being a 12-month fixture from western Washington down through western California and southern Arizona, and the Buff-bellied Hummingbird resides along the Gulf Coast during winter, along with several species of western hummingbirds (especially Rufous) that show up in limited numbers the South during winter. In these southern latitudes, many locations maintain feeders year-round.

Planning a hummingbird-viewing trip requires that you consider the yearly schedule of the birds; again, one of the best places to do that is the Cornell Lab of Ornithology's eBird, which allows users to report their bird sightings and view sightings recorded by other people. eBird has many other search parameters and information interfaces—it's a fun and educational website.

In many ways, hummingbirds are portals into the wider natural world, and a visit to a refuge, nature center, or even a remote inn or café that caters to hummingbirds frequently presents broader opportunities to enjoy many facets of nature. The following state-by-state list of locations that maintain hummingbird feeders for public view is dominated by state and federal properties; most offer far more than just the birdfeeders, so be sure to research your destination and plan your trip to accordingly. These sites preserve special places—historic locations, natural wonders, intriguing and sometimes imperiled habitats. And some maintain both hummingbird feeders and pollinator gardens; a few have only pollinator gardens, with flowers that attract hummingbirds to the site.

RULES FOR THE ROAD

Follow these simple rules to get the most out of your road trip to see hummingbirds.

Always call ahead. Before you hit the road, call the site you want to visit to make sure the hummingbird feeders are being maintained and visited by hummingbirds. Doublecheck open hours and ask about the best times to see the hummingbirds.

Leave pets at home. In almost all cases, publicly accessible birding sites do not allow pets, and because in most places hummingbirds are birds of summer, warm weather means unsafe conditions for leaving Fido in the vehicle.

Leave a little green. Some hummingbird viewing sites ask for donations in lieu of entry fees, and these funds go to maintaining the grounds and feeders, so be generous.

Stay for lunch. If you visit businesses that maintain hummingbird feeders for the enjoyment of customers, be a patron—have a meal, book a room, buy supplies.

Keep it quiet. Even though the birds themselves may be well conditioned to human onlookers, keep a low voice at wildlife-viewing areas in deference to other visitors.

Express gratitude. Many hummingbird feeder stations are maintained by volunteers or by staff members with other duties. Say thanks when you leave and send a note of thanks when you get home.

VIEWING HUMMINGBIRDS, STATE BY STATE

ALABAMA

5 Rivers–Alabama Delta Resource Center
outdooralabama.com/activities/5-rivers-alabama-delta-resource-center

Bon Secour National Wildlife Refuge
fws.gov/refuge/Bon_Secour

Dauphin Island Audubon Bird Sanctuary
dauphinisland.org/audubon-bird-sanctuary

Fort Gaines Historic Site
dauphinisland.org/fort-gaines

Fort Morgan Historic Site
fort-morgan.org

Wheeler National Wildlife Refuge
fws.gov/refuge/wheeler
friendsofwheelerrefuge.com

ALASKA

Alaska Wildlife Conservation Center
alaskawildlife.org

Annie Mae Lodge
anniemae.com

George Inlet Lodge
catchcrabs.com

Portage Glacier Lodge Café
Girdwood / (907) 783-3117

Taku Glacier Lodge
wingsairways.com

ARIZONA

American Museum of Natural History Southwestern Research Station
amnh.org/research/southwestern-research-station

Arizona-Sonora Desert Museum
desertmuseum.org

Ash Canyon Bird Sanctuary
Hereford / (520) 432-1388

Battiste's Bed, Breakfast, and Birds
Hereford / (520) 335-1030
Day birders must call ahead for reservations; donations encouraged.

Beatty's Guest Ranch
beattysguestranch.com

Boyce Thompson Arboretum
btarboretum.org

Buenos Aires National Wildlife Refuge
fws.gov/refuge/buenos_aires

Cabeza Prieta National Wildlife Refuge
fws.gov/refuge/cabeza_prieta

Carr Canyon, Coronado National Forest
fs.usda.gov/recarea/coronado/recarea/?recid=25476

Catalina State Park
azstateparks.com/catalina

Cave Creek Canyon
friendsofcavecreekcanyon.com

George Walker House
thegeorgewalkerhouse.com

Imperial National Wildlife Refuge
fws.gov/refuge/imperial

Madera Canyon
friendsofmaderacanyon.org

Mill Restaurant, Prescott National Forest
Crown King / (928) 632-7133

Montezuma Castle National Monument
nps.gov/moca

Patagonia Lake State Park
azstateparks.com/patagonia-lake

Patagonia–Sonoita Creek Preserve
nature.org/en-us/get-involved/how-to-help/places-we-protect/patagonia-sonoita-creek-preserve

Paton Center for Hummingbirds
tucsonaudubon.org/go-birding/tucson-audubons-paton-center-for-hummingbirds

Ramsey Canyon Preserve
nature.org/en-us/get-involved/how-to-help/places-we-protect/ramsey-canyon-preserve

Sipe White Mountain Wildlife Area
Springerville / (928) 532-3680

Southeastern Arizona Bird Observatory
sabo.org

Wild Outdoor World of Arizona
wowarizona.org

ARKANSAS

Fred Berry Conservation Education Center
agfc.com/en/explore-outdoors/nature-and-education-centers/fbcec

Gaston's White River Resort
gastons.com

Governor Mike Huckabee Delta Rivers Nature Center
agfc.com/en/explore-outdoors/nature-and-education-centers/gmhdrnc

Hobbs State Park–Conservation Area
arkansasstateparks.com/parks/hobbs-state-park-conservation-area

Holla Bend National Wildlife Refuge
fws.gov/refuge/holla_bend

Mississippi River State Park
arkansasstateparks.com/parks/mississippi-river-state-park

Mount Magazine State Park
arkansasstateparks.com/parks/mount-magazine-state-park

Cardinal flowers at Pinnacle Mountain State Park are magnets for hummingbirds and butterflies.

Pinnacle Mountain State Park
arkansasstateparks.com/parks/pinnacle-mountain-state-park

Potlatch Conservation Education Center
agfc.com/en/explore-outdoors/nature-and-education-centers/pcec

CALIFORNIA

Audubon Center at Debs Park
debspark.audubon.org

Audubon Kern River Preserve
kern.audubon.org

Big Morongo Canyon Preserve
bigmorongo.org

El Dorado Nature Center
longbeach.gov/park/park-and-facilities/parks-centers-pier/el-dorado-nature-center

Living Desert Zoo and Gardens
livingdesert.org

Madrona Marsh Preserve
friendsofmadronamarsh.com

Sacramento National Wildlife Refuge
fws.gov/refuge/Sacramento

San Diego Botanic Garden
sdbgarden.org

San Diego Zoo Hummingbird House
zoo.sandiegozoo.org

Stone Lakes National Wildlife Refuge
fws.gov/refuge/stone_lakes

Tucker Wildlife Sanctuary
tuckerwildlife.org

Tule Lake National Wildlife Refuge
fws.gov/refuge/Tule_Lake/About_the_Complex.html

UC Santa Cruz Arboretum and Botanic Garden
arboretum.ucsc.edu/index.html

COLORADO

Arapaho National Wildlife Refuge
fws.gov/refuge/arapaho

Audubon Center at Chatfield State Park
denveraudubon.org/audubоncenter
Bear Creek Nature Center
communityservices.elpasoco.com/nature-centers/bear-creek-nature-center

Wild irises illuminate a verdant meadow at Arapaho National Wildlife Refuge.

Echo Lake Lodge
echolakelodgeco.com

Mishawaka Restaurant
themishawaka.com

Potbelly Restaurant & Lounge
Red Feather Lakes / (970) 881-2984

Starsmore Visitor and Nature Center
coloradosprings.gov/parks/page/starsmore-visitor-and-nature-center

Stonewall Shopping Bag
stonewallshoppingbag.com

Trading Post Resort
tradingpostresort.com

Vail Nature Center
walkingmountains.org/locations/vail-nature-center

CONNECTICUT

Bent of the River Audubon Center
bentoftheriver.audubon.org

Connecticut Audubon Center at Pomfret
ctaudubon.org/pomfret

Cove Island Park
ct.audubon.org/conservation/cove-island-park

Deer Pond Farm Sanctuary
ctaudubon.org/deer-pond-farm-home/

Greenwich Audubon Center
greenwich.audubon.org

Lighthouse Point Park
ct.audubon.org/conservation/lighthouse-point-park

Milford Point
ctaudubon.org/coastal-center-home

DELAWARE

Ashland Nature Center
delawarenaturesociety.org/centers/ashland-nature-center

Bombay Hook National Wildlife Refuge
fws.gov/refuge/Bombay_Hook

Middle Run Natural Area
delawarenaturesociety.org/centers/middle-run-natural-area

FLORIDA

Bok Tower Gardens
boktowergardens.org

Castellow Hammock Preserve and Nature Center
miamidade.gov/parks/castello-hammock.asp

Corkscrew Swamp Sanctuary
corkscrew.audubon.org

Fairchild Tropical Botanic Garden
fairchildgarden.org

Fort De Soto County Park
pinellascounty.org/park/05_ft_desoto.htm

Lake Apopka Wildlife Drive
sjrwmd.com/lands/recreation/lake-apopka/wildlife-drive

Lake Lotus Park
altamonte.org/959/Lake-Lotus-Park

Mead Botanical Garden
meadgarden.org

Merritt Island National Wildlife Refuge
fws.gov/refuge/Merritt_Island

Orlando Wetlands Park
cityoforlando.net/wetlands/plan-your-visit/

St. Marks National Wildlife Refuge
fws.gov/refuge/st_marks
stmarksrefuge.org

GEORGIA

Atlanta Botanical Garden
atlantabg.org

Blue Heron Nature Preserve
bhnp.org

Clyde Shepherd Nature Preserve
cshepherdpreserve.org

Harris Neck National Wildlife Refuge
fws.gov/refuge/harris_neck

Jekyll Island Campground Bird Sanctuary
jekyllisland.com/lodging/jekyll-island-campground

Newman Wetlands Center
ccwa.us/newman-wetlands-center

Rum Creek Wildlife Management Area
georgiawildlife.com/rum-creek-wma

Smith-Gilbert Gardens
smithgilbertgardens.com

State Botanical Garden of Georgia
botgarden.uga.edu

IDAHO

Brockman's Hummingbird Feeding Station
idfg.idaho.gov/ifwis/ibt/site.aspx?id=SW39#maincontent

Camas National Wildlife Refuge
fws.gov/refuge/camas
friendsofcamas.org

Kelly Forks Campground
fs.usda.gov/recarea/nezperceclearwater/recarea/?recid=80050

Kootenai National Wildlife Refuge
fws.gov/refuge/kootenai

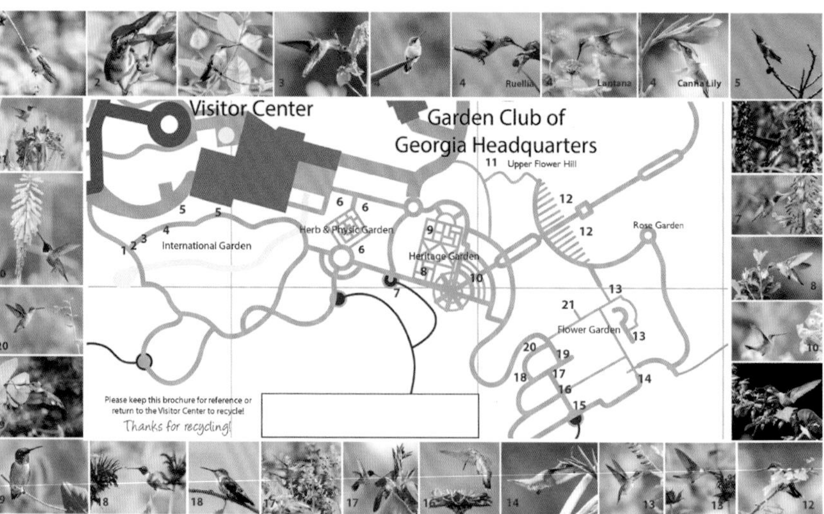

The Hummingbird Trail at the State Botanical Garden of Georgia features a veritable who's-who of the best flowers for attracting hummingbirds.

Lolo Pass Visitor Center
fs.usda.gov/recarea/nezperceclearwater/recarea/?recid=80112

Silver Creek Preserve
nature.org/en-us/get-involved/how-to-help/places-we-protect/silver-creek-preserve

The Springs Hot Springs Retreat
thespringsid.com

Wilderness Gateway Campground
fs.usda.gov/recarea/nezperceclearwater/recarea/?recid=80725

ILLINOIS

Ballard Nature Center
ballardnaturecenter.org

Cache River Wetlands Center
friendsofthecache.org/wetlands-center

Fullersburg Woods Nature Education Center
dupageforest.org/fullersburg-woods-nature-education-center

Isle a la Cache Museum
reconnectwithnature.org/preserves-trails/visitor-centers/isle-a-la-cache-museum

Little Red Schoolhouse Nature Center
fpdcc.com/places/locations/little-red-schoolhouse-nature-center

Plum Creek Nature Center
reconnectwithnature.org/preserves-trails/visitor-centers/plum-creek-nature-center

Sagawau Environmental Learning Center
fpdcc.com/places/locations/sagawau-environmental-learning-center

Sand Ridge Nature Center
fpdcc.com/places/locations/sand-ridge-nature-center

Severson Dells Nature Center
seversondells.com

Starved Rock State Park
starvedrockstatepark.org/start/the-visitor-center

Volo Bog State Natural Area
www2.illinois.gov/dnr/Parks/Pages/VoloBog.aspx
friendsofvolobog.org

War Bluff Valley Sanctuary
illinoisaudubon.org/location/war-bluff-sanctuary

INDIANA

Eagle Creek Ornithology Center
eaglecreekpark.org/ornithology-center

Holliday Park Nature Center
hollidaypark.org

Indiana Dunes State Park Nature Center
indianadunes.com/eat-play-shop/recreation/indiana-dunes-state-park-nature-center

Mary Gray Bird Sanctuary
indianaaudubon.org/mgbs

Scales Lake Park Gatehouse
warrickcountyparks.com/scales-lake-park

Turkey Run State Park Nature Center
turkeyrunstatepark.com/nature_center

IOWA

F. W. Kent Park
mycountyparks.com/County/Johnson/Park/F-W-Kent-Park.aspx

Hartman Reserve Nature Center
hartmanreserve.org

Lake Meyer Park
winneshiekwild.com
mycountyparks.com/County/Winneshiek/Park/Lake-Meyer-Park-and-Campground.aspx

Pikes Peak State Park
iowadnr.gov/Places-to-Go/State-Parks/Iowa-State-Parks/parkdetails/parkid/610141

KANSAS

Chaplin Nature Center
wichitaaudubon.org

A young onlooker watches a licensed hummingbird bander fit a Ruby-throated Hummingbird with a tiny leg band during an event at Land Between the Lakes Woodlands Nature Station.

Ernie Miller Park
erniemiller.com

Great Plains Nature Center
gpnc.org

Kansas Wetlands Education Center
wetlandscenter.fhsu.edu

Southeast Kansas Nature Center
ksoutdoors.com/KDWPT-Info/Locations/Museums-and-Nature-Centers/Southeast-Kansas-Nature-Center

KENTUCKY

Campbell County Environmental Education Center
campbell.ca.uky.edu/sites/campbell.ca.uky.edu/files/eec_brochure_for_new_website.pdf

Clarks River National Wildlife Refuge
fws.gov/refuge/clarks_river

John James Audubon State Park
parks.ky.gov/parks/recreationparks/john-james

Land Between the Lakes Woodlands Nature Station
landbetweenthelakes.us/seendo/attractions/nature-station

Natural Bridge State Resort Park
parks.ky.gov/parks/resortparks/natural-bridge

Salato Wildlife Center
fw.ky.gov/Education/Pages/Salato-Wildlife-Education-Center.aspx

LOUISIANA

Acadiana Park Nature Station
naturestation.org

Black Bayou Lake National Wildlife Refuge
fws.gov/refuge/Black_Bayou_Lake

Bluebonnet Swamp Nature Center
brec.org/index.cfm/park/BluebonnetSwamp

Cypress Island Preserve
nature.org/en-us/get-involved/how-to-help/places-we-protect/cypress-island

Red River National Wildlife Refuge
fws.gov/refuge/Red_River
friendsofredriver.org

St. Catherine Creek National Wildlife Refuge
fws.gov/refuge/st_catherine_creek

MAINE

Asticou Azalea Garden
gardenpreserve.org

Borestone Mountain Audubon Sanctuary
maineaudubon.org/visit/borestone

Camden Hills State Park
maine.gov/dacf/parks/trail_activities/camden_trail_conditions.shtml

Coastal Maine Botanical Gardens
mainegardens.org

Gilsland Farm Audubon Center
maineaudubon.org/visit/gilsland-farm

Wells Reserve at Laudholm
wellsreserve.org

MARYLAND

Blackwater National Wildlife Refuge
fws.gov/refuge/blackwater

Patterson Park Audubon Center
patterson.audubon.org

Patuxent Research Refuge
fws.gov/refuge/Patuxent

Pickering Creek Audubon Center
pickering.audubon.org

Piney Run Park
pineyrunpark.org

MASSACHUSETTS

Blue Hills Trailside Museum
massaudubon.org/get-outdoors/wildlife-sanctuaries/blue-hills-trailside-museum

Felix Neck Wildlife Sanctuary
massaudubon.org/get-outdoors/wildlife-sanctuaries/felix-neck

Ipswich River Wildlife Sanctuary
massaudubon.org/ipswichriver

Monomoy National Wildlife Refuge
fws.gov/refuge/Monomoy/

MICHIGAN

Chippewa Nature Center
chippewanaturecenter.org

Harris Nature Center
meridian.mi.us/visitors/hidden-gems/visit-harris-nature-center

Kalamazoo Nature Center
naturecenter.org

Lake Erie Metropark
metroparks.com/parks/lake-erie-metropark

Lake St. Clair Nature Center
metroparks.com/facilities-education/lake-st-clair-metropark-nature-center

Sarett Nature Center
sarett.com

Tawas Point State Park
michigandnr.com/ParksandTrails/Details.aspx?id=499&type=SPRK

MINNESOTA

Afton State Park
dnr.state.mn.us/state_parks/park.html?id=spk00100#homepage

Hawk Ridge Bird Observatory
hawkridge.org

Henderson Hummingbird Garden at Bender Park
henderson-mn.com/visitors.html

Itasca State Park
dnr.state.mn.us/state_parks/park.html?id=spk00181#homepage

Ney Nature Center
neycenter.org

Quarry Hill Nature Center
qhnc.org

Tamarac National Wildlife Refuge
fws.gov/refuge/Tamarac
tamaracfriends.org/visitor-center

Westwood Hills Nature Center
stlouispark.org/government/departments-divisions/parks-rec/westwood-hills-nature-center/-selcat-2

Whitewater State Park
dnr.state.mn.us/state_parks/park.html?id=spk00280#homepage

ABOVE Featuring feeders and flowers, the Pascagoula River Audubon Center attracts Ruby-throated Hummingbirds, especially during the spring and fall migrations.

BELOW Missouri's beautiful Twin Pines Conservation Education Center sits just off Highway 60, about a mile southeast of Winona.

MISSISSIPPI

Mississippi Museum of Natural Science
mdwfp.com/museum.aspx

Pascagoula River Audubon Center
pascagoula.audubon.org

Sam D. Hamilton Noxubee National Wildlife Refuge
fws.gov/refuge/Noxubee

Strawberry Plains Audubon Center
strawberry.audubon.org

MISSOURI

Burr Oak Woods Conservation Area
nature.mdc.mo.gov/discover-nature/places/burr-oak-woods-ca

Cape Girardeau Conservation Nature Center
nature.mdc.mo.gov/discover-nature/places/cape-girardeau-conservation-nature-center

Gorman Conservation Discovery Center
nature.mdc.mo.gov/discover-nature/ places/gorman-anita-b-conservation-discovery-center

Loess Bluffs National Wildlife Refuge
fws.gov/refuge/Loess_Bluffs

Missouri Department of Conservation, Southeast and St. Louis Regional Offices
mdc.mo.gov/contact-engage/regional-offices

Powder Valley Conservation Nature Center
nature.mdc.mo.gov/discover-nature/places/powder-valley-cnc

Runge Nature Center
nature.mdc.mo.gov/discover-nature/places/runge-nature-center

Springfield Conservation Nature Center
nature.mdc.mo.gov/discover-nature/places/springfield-cnc

Twin Pines Conservation Education Center
nature.mdc.mo.gov/discover-nature/places/twin-pines-conservation-education-center

MONTANA

Blacktail Guest Ranch
blacktailranch.com

Diamond R Guest Ranch
diamondrranch.com

Lee Metcalf National Wildlife Refuge
fws.gov/refuge/lee_metcalf

Skalkaho Steak House
Hamilton / (406) 363-3522

Smoky Bear Ranch
smokybear.com

Tizer Botanic Gardens and Arboretum
tizergardens.com

NEBRASKA

Eugene T. Mahoney State Park
outdoornebraska.gov/mahoney

Indian Cave State Park
outdoornebraska.gov/indiancave

Lewis and Clark State Recreation Area
outdoornebraska.gov/lewisandclark

Ponca State Park
outdoornebraska.gov/ponca

Rock Creek Station State Historical Park
outdoornebraska.gov/rockcreekstation

Wildcat Hills Nature Center
outdoornebraska.gov/wildcathills

NEVADA

Desert National Wildlife Refuge
fws.gov/refuge/Desert

Discovery Park
redrockaudubon.com

Henderson Bird Viewing Preserve
cityofhenderson.com/henderson-happenings/facilities/henderson-bird-viewing-preserve

Moapa Valley National Wildlife Refuge
fws.gov/refuge/moapa_valley/

Red Rock Canyon National Conservation Area
blm.gov/programs/national-conservation-lands/nevada/red-rock-canyon
redrockcanyonlv.org

NEW HAMPSHIRE

Weeks State Park
nhstateparks.org/visit/state-parks/weeks-state-park

NEW JERSEY

Cape May Bird Observatory's Center for Research and Education
njaudubon.org/venue/cape-may-bird-observatorys-center-for-research-and-education

Edwin B. Forsythe National Wildlife Refuge
fws.gov/refuge/edwin_b_forsythe
friendsofforsythe.org

Great Swamp National Wildlife Refuge
fws.gov/refuge/great_swamp

James A. McFaul Environmental Center
co.bergen.nj.us/ja-mcfaul-environmental-center

Manasquan Reservoir Environmental Center
monmouthcountyparks.com/page.aspx?ID=2627

Wallkill River National Wildlife Refuge
fws.gov/refuge/wallkill_river

NEW MEXICO

Bear Mountain Lodge
bearmountainlodge.com

Bosque del Apache National Wildlife Refuge
fws.gov/refuge/bosque_del_apache

Capulin Volcanic National Monument
nps.gov/cavo/index.htm

Chiricahua Desert Museum
chiricahuadesertmuseum.com

Dripping Springs Natural Area
blm.gov/visit/dripping-springs-natural-area

El Malpais National Monument
nps.gov/elma/index.htm

Mesilla Valley Bosque State Park
emnrd.state.nm.us/SPD/mesillavalleystatepark.html

Randall Davey Audubon Center
randalldavey.audubon.org

Rio Grande Nature Center State Park
emnrd.state.nm.us/spd/riograndenaturecenterstatepark.html

Sevilleta National Wildlife Refuge
fws.gov/refuge/sevilleta

Sugarite Canyon State Park
emnrd.state.nm.us/SPD/sugaritecanyonstatepark.html

Tererro General Store
pecoswilderness.com

NEW YORK

Adirondack Interpretive Center
esf.edu/aic/

Audubon Community Nature Center
auduboncnc.org

Beaver Meadow Audubon Center
buffaloaudubon.org/beaver-meadow-audubon-center.html

Bedford Audubon
bedfordaudubon.org

Long Island National Wildlife Refuge Complex
fws.gov/northeast/longislandrefuges/

Marshlands Conservancy
parks.westchestergov.com/marshlands-conservancy

Montezuma Audubon Center
ny.audubon.org/montezuma

Montezuma National Wildlife Refuge
fws.gov/refuge/Montezuma
friendsofmontezuma.org

Reinstein Woods Nature Preserve
reinsteinwoods.org

Sonnenberg Gardens and Mansion State Historic Park
sonnenberg.org

Tifft Nature Preserve
tifft.org

NORTH CAROLINA

Airlie Gardens
airliegardens.org

Big Bloomers Flower Farm
bigbloomersflowerfarm.com

Daniel Stowe Botanical Garden
dsbg.org

Elizabethan Gardens
elizabethangardens.org

Howell Woods Environmental Learning Center
johnstoncc.edu/howellwoods

Kaleideum North Outdoor Park
north.kaleideum.org

Mattamuskeet National Wildlife Refuge
fws.gov/refuge/mattamuskeet

North Carolina Botanical Garden
ncbg.unc.edu

Pilot Mountain State Park
ncparks.gov/pilot-mountain-state-park

Getting a closeup view of the flora at New York's Tifft Nature Preserve.

Red Oak Brewery
redoakbrewery.com

Reedy Creek Nature Center and Preserve
mecknc.gov/ParkandRec/StewardshipServices/Nature-Centers/Pages/Reedy.aspx

Riverbend Park
catawbacountync.gov/county-services/parks/riverbend-park

Salem Lake Park
cityofws.org/Departments/Recreation-Parks/Salem-Lake

A male Ruby-throated Hummingbird at the Black Swamp Bird Observatory, Magee Marsh Wildlife Area.

Weymouth Woods Sandhills Nature Preserve
ncparks.gov/weymouth-woods-sandhills-nature-preserve

Wright's Birding Center
wrightsbirdingcenter.wordpress.com

NORTH DAKOTA

White Horse Hill National Game Preserve
fws.gov/refuge/White_Horse_Hill

OHIO

Black Swamp Bird Observatory
bsbo.org

Cincinnati Nature Center
cincynature.org

Ottawa National Wildlife Refuge
fws.gov/refuge/ottawa
friendsofottawanwr.org

Trautman Nature Center at Maumee Bay State Park
maumeebaystatepark.org/NatureCenter.html

Vermilion River Reservation
loraincountymetroparks.com/vermilion-river-reservation

OKLAHOMA

Oxley Nature Center at Redbud Valley Nature Preserve
oxleynaturecenter.org

Salt Plains National Wildlife Refuge
fws.gov/refuge/salt_plains

Tulsa Botanic Garden
tulsabotanic.org

Washita National Wildlife Refuge
fws.gov/refuge/Washita

Wichita Mountains Wildlife Refuge
fws.gov/refuge/Wichita_Mountains

OREGON

Audubon Society of Portland Nature Sanctuary
audubonportland.org/sanctuaries/visiting

Malheur National Wildlife Refuge
fws.gov/refuge/malheur

Silver Falls State Park
oregonstateparks.org/index.cfm?do=parkPage.dsp_parkPage&parkId=151

William L. Finley National Wildlife Refuge
fws.gov/refuge/william_l_finley

Pollinator garden at Rhode Island's Norman Bird Sanctuary.

PENNSYLVANIA

Clarence Schock Memorial Park at Governor Dick
parkatgovernordick.org

Hawk Mountain Sanctuary
hawkmountain.org

Hershey Gardens
hersheygardens.org

Longwood Gardens
longwoodgardens.org

Middle Creek Wildlife Management Area
pgc.pa.gov/InformationResources/AboutUs/ContactInformation/Southeast/MiddleCreekWildlifeManagementArea/Pages/default.aspx

Ned Smith Center
nedsmithcenter.org

Peace Valley Nature Center
peacevalleynaturecenter.org

Pennypack Ecological Restoration Trust
pennypacktrust.org

Shaver's Creek Environmental Center
shaverscreek.org

Wildwood Park
wildwoodlake.org

RHODE ISLAND

Norman Bird Sanctuary
normanbirdsanctuary.org

SOUTH CAROLINA

Barnwell State Park
southcarolinaparks.com/barnwell

Beidler Forest Audubon Center
sc.audubon.org/visit/beidler

Devils Fork State Park
southcarolinaparks.com/devils-fork

Edisto Beach State Park
southcarolinaparks.com/edisto-beach

Hunting Island State Park
southcarolinaparks.com/hunting-island

Huntington Beach State Park
southcarolinaparks.com/huntington-beach

Lee State Park
southcarolinaparks.com/lee

Magnolia Plantation and Gardens
magnoliaplantation.com

Musgrove Mill State Historic Site
southcarolinaparks.com/musgrove-mill

Myrtle Beach State Park
southcarolinaparks.com/myrtle-beach

Santee State Park
southcarolinaparks.com/santee

Table Rock State Park
southcarolinaparks.com/table-rock

SOUTH DAKOTA

Good Earth State Park
gfp.sd.gov/parks/detail/good-earth-state-park

Oakwood Lakes State Park
gfp.sd.gov/parks/detail/
oakwood-lakes-state-park

TENNESSEE

Cross Creeks National Wildlife Refuge
fws.gov/refuge/Cross_Creeks

Ijams Nature Center
ijams.org

Lichterman Nature Center
memphismuseums.org/
lichterman-nature-center

Meeman-Shelby Forest State Park
tnstateparks.com/parks/meeman-shelby

Reflection Riding Arboretum
reflectionriding.org

Steele Creek Park
bristoltn.org/540/Steele-Creek-Nature-Center
friendsofsteelecreek.org

Tennessee National Wildlife Refuge
fws.gov/refuge/Tennessee

Warner Park
nashville.gov/Parks-and-Recreation/
Nature-Centers-and-Natural-Areas/
Warner-Park-Nature-Center.aspx

TEXAS

Bentsen–Rio Grande Valley State Park
theworldbirdingcenter.com/Bentsen.html#

Big Bend National Park
nps.gov/bibe/index.htm

Chihuahuan Desert Nature Center
cdri.org

Christmas Mountains Oasis
cmoasis.blogspot.com
Private property; email owner (see website) before visiting; donations encouraged.

Davis Mountains State Park
tpwd.texas.gov/state-parks/davis-mountains

The Fossil Discovery Exhibit is one of many popular stops at sprawling Big Bend National Park, where hummingbirds abound.

Estero Llano Grande State Park
tpwd.texas.gov/state-parks/estero-llano-grande

Fennessey Ranch
fennesseyranch.com

Hotel Limpia
hotellimpia.com

Hummer House
hummerhouse.com

Kleb Woods Nature Preserve and Center
pct3.com/Parks/
Kleb-Woods-Nature-Preserve-Center

Quinta Mazatlan World Birding Center
quintamazatlan.com

Sabal Palm Sanctuary
sabalpalmsanctuary.org

Santa Ana National Wildlife Refuge
fws.gov/refuge/santa_ana

Texas Chenier Plain National Wildlife Refuge Complex
fws.gov/refuge/Anahuac/
friendsofanahuacnwr.com

UTAH

Antelope Island State Park
stateparks.utah.gov/parks/antelope-island

Best Friends Animal Sanctuary
bestfriends.org/sanctuary/
visit-our-utah-sanctuary

Red Canyon Lodge
redcanyonlodge.com

Stokes Nature Center
logannature.org

Tonaquint Nature Center
sgcity.org/sportsandrecreation/
recreationfacilities/tonaquintnaturecenter

VERMONT

Birds of Vermont Museum
birdsofvermont.org

Grand Isle State Park
vtstateparks.com/grandisle.html

Green Mountain Audubon Center
vt.audubon.org/about-us/
green-mountain-audubon-center

Jamaica State Park
vtstateparks.com/jamaica.html

Ricker Pond State Park
vtstateparks.com/ricker.html

Smugglers' Notch State Park
vtstateparks.com/smugglers.html

Townshend State Park
vtstateparks.com/townshend.html

Vermont Institute of Natural Science
vinsweb.org

VIRGINIA

Blandy Experimental Farm
blandy.virginia.edu/home

Chippokes Farm and Forestry Museum
dcr.virginia.gov/state-parks/
chippokes-farm-museum

Fairy Stone State Park
dcr.virginia.gov/state-parks/
fairy-stone#general_
information

Checking out the birds at the wonderful Green Mountain Audubon Center.

The Brinton Museum is housed in a beautiful historic ranch house.

Huntley Meadows Park
fairfaxcounty.gov/parks/huntley-meadows
friendsofhuntleymeadows.org

Mountain Lake Lodge
mtnlakelodge.com

Widewater State Park
dcr.virginia.gov/state-parks/widewater#general_information

WASHINGTON

Lake Quinault Lodge, Olympic National Park
olympicnationalparks.com/lodging/lake-quinault-lodge

Little Pend Oreille National Wildlife Refuge
fws.gov/refuge/little_pend_oreille

McNary National Wildlife Refuge
fws.gov/refuge/McNary

WEST VIRGINIA

Cranberry Mountain Nature Center
fs.usda.gov/recarea/mnf/recarea/?recid=7049

WISCONSIN

Crex Meadows Wildlife Area
dnr.wi.gov/topic/lands/WildlifeAreas/crex.html
crexmeadows.org

Hunt Hill Audubon Sanctuary
hunthill.org

Necedah National Wildlife Refuge
fws.gov/refuge/necedah

Schlitz Audubon Nature Center
schlitzaudubon.org

Wyalusing State Park
dnr.wi.gov/topic/parks/name/wyalusing

WYOMING

Brinton Museum
thebrintonmuseum.org

Craig Thomas Discovery and Visitor Center, Grand Teton National Park
nps.gov/grte/planyourvisit/visitorcenters.htm

Shell Falls Interpretive Site
fs.usda.gov/recarea/bighorn/recarea/?recid=30868

HUMMINGBIRD FESTIVALS

HUMMINGBIRDS ARE SO UNIVERSALLY POPULAR that they inspire special events dedicated to them around the country. These educational festivals range from multiday celebrations to one-day events, and many of them are scheduled for migration times during spring or late summer so that licensed hummingbird banders can capture and band birds—to the delight of onlookers—while explaining the purpose and value of the data they collect. Most hummingbird festivals cater to all ages and feature family-friendly activities and fun educational events for kids. The best known require advance registration, testifying to their popularity; smaller one-day events require only that participants arrive eager to learn. The list that follows details most ongoing annual hummingbird festivals.

ALASKA

Alaska Hummingbird Festival / April
alaskacenters.gov/visitors-centers/ketchikan

ARIZONA

High Country Hummingbird Festival / late July
azgfd.com/wildlife/viewing/events/hummingbird-festival

Sedona Hummingbird Festival / early August
hummingbirdsociety.org

CALIFORNIA

Hummingbird Day / mid-May
arboretum.ucsc.edu

COLORADO

North Cheyenne Cañon Hummingbird Festival / early May
cheyennecanon.org/events/hummingbird-festival

IDAHO

Rudeen Ranch Hummingbird Roundup / early summer
hummingbirdroundup.com

ILLINOIS

Hummingbird Fest / mid-August
fpdcc.com/places/locations/sagawau-environmental-learning-center

Hummingbird Festival and Pollination Celebration / late August, early September
sugargrovenaturecenter.org

INDIANA

Indiana Audubon Hummingbird Migration Celebration / mid-August
indianaaudubon.org

KENTUCKY

Woodlands Nature Station Annual Hummingbird Festival / first weekend in August
landbetweenthelakes.us/seendo/attractions/nature-station

LOUISIANA

Feliciana Hummingbird Celebration / mid-September
stfrancisvillefestivals.com/features/feliciana-hummingbird-celebration

MINNESOTA

Henderson Hummingbird Hurrah / late August
hendersonhummingbirdhurrah.com

MISSISSIPPI

Hummingbird Migration and Nature Celebration / early September
strawberry.audubon.org/hummingbird

NEW MEXICO

Hummingbird Festival / late July
mimbrescultureheritagesite.org

NORTH CAROLINA

Big Bloomers Hummingbird Day / late July
bigbloomersflowerfarm.com

Hummingbird Festival at Reedy Creek / August
mecknc.gov/ParkandRec/StewardshipServices/NatureCenters/Pages/Reedy.aspx

Wild Wings! / early August
dsbg.org

TENNESSEE

Wonder of Hummingbirds Festival / late August
ijams.org

TEXAS

Davis Mountains Hummingbird Celebration / late August
davismountainshummingbirdcelebration.com

Rockport-Fulton HummerBird Celebration / late September
rockport-fulton.org/HB

SIX

Hummingbirds Abroad: A Gallery of Species

Ecotourism has become a driving economic force worldwide, and in the Americas, hummingbirds deserve some of the credit for bringing a significant new form of growth to communities in several nations. Certainly ecotourism has its downsides, such as the potential degradation of fragile ecosystems (now threatened by being loved to death, so to speak), but the benefits of ecotourism usually outweigh the negative consequences. Ecotourism promotes meaningful interaction and mutual understanding between geographically, politically, and linguistically disparate cultures. In some places, tourists motivated by experiencing pristine nature provide economic stimuli to traditionally and sometimes chronically poor communities and societies; in others, ecotourism provides a profitable alternative to extractive industries, spurring preservation or restoration of threatened ecosystems.

Purple-throated Woodstar.

In the Americas—especially Central America and South America—birds are a significant tourist draw, supporting a robust ecotourism industry. Brazil, for example, the largest nation in South America, is home to more than 1,813 species of birds. Far smaller in total land mass, Peru has 1,858 species of birds, and Colombia is the world champion, with 1,878 species of birds. Even diminutive Costa Rica, a popular destination for ecotourists and a country smaller than West Virginia, has 850 bird species, nearly as many as the entire United States. Among these many hundreds of Central and South American bird species are about 330 different hummingbirds. Ecuador alone has 163 hummingbird species—no wonder it and most other South American and Central American nations are so popular with traveling hummingbird enthusiasts.

Birdwatchers flock to well-known hummingbird hotspots and destinations recently made accessible. Ecolodges and local guides cater expertly to hummingbird enthusiasts; hummingbird feeding stations are common at Central and South American ecolodges, preserves, and reserves, some privately owned and some managed by local or national governments. With expertise in various nations and regions therein, birding tour organizers in the United States and throughout the Americas create itineraries for expeditions that take all the guesswork out of traveling to other countries; the top tour operators are the best means by which to visit several hotspots, for the best chance of seeing many different mesmerizing hummingbird species.

ABOVE Throughout the Americas, as here in Ecuador, many communities are keenly aware of the significance of hummingbirds to local ecotourism.

BELOW Ecolodges, such as Guango Lodge in Ecuador, obviously and expertly cater to hummingbird fans.

Anyone intrigued by hummingbirds would revel in a well-organized guided expedition to the bird-rich nations of Central and South America. Planning such a trip requires time and foresight, and specialty tour operators help clients with all the travel particulars for specific countries,

Professional birding tour operators, working with local lodges and guides, provide a great way to visit hummingbird hotspots throughout the Americas.

including paperwork and immunization requirements and safety considerations. Once on the ground in a new country, in a destination where the cacophony of the jungle promises many dozens if not hundreds of exotic birds you've never seen before, the onus is on you to be a conscientious traveler, a gracious guest in a land with customs different from your own.

For many people, those faraway hummingbird paradises are dream destinations, places we may never get to visit. But we can still deeply appreciate the amazing world of hummingbirds and marvel at their seemingly endless variety. For the dreamers among us, the following gallery is but a teaser, a sampling of the hundreds of species of these beguiling creatures.

The Amethyst Woodstar (*Calliphlox amethystina*) is widespread in South America, though absent from the central and upper Amazon Basin.

The tiny Bee Hummingbird (*Mellisuga helenae*), endemic to Cuba, is the world's smallest bird

The handsome Black Jacobin (*Florisuga fusca*) is a resident of the Atlantic Forest from Brazil south to northeastern Argentina.

This male Black-throated Mango (*Anthracothorax nigricollis*) proves that, at least for brief stints, hummingbirds can fly upside down.

The Black-throated Mango occurs from Panama south to Argentina and also on Trinidad and Tobago. The female, shown here, is striking.

The Blue-chested Hummingbird (*Amazilia amabilis*), a Central American species, is found on the Atlantic slope of Nicaragua and Costa Rica, south through Panama and the Pacific slope of Colombia and Ecuador.

Though tiny, the male Booted Racket-tail (*Ocreatus underwoodii*) is spectacular.

Booted Racket-tails are mountain specialists, living primarily at mid-elevations of the Andes. The female, shown here, differs dramatically from the resplendent male.

The colorful Brazilian Ruby (*Clytolaema rubricauda*) lives only in Brazil, where it inhabits the Atlantic Forest near the coast and at elevations up to 6,000 feet.

The Buffy Helmetcrest (*Oxypogon stuebelii*) is found only in the Andean páramo (high alpine tundra) of central Colombia.

The Chestnut-breasted Coronet (*Boissonneaua matthewsii*) is a denizen of the Andean cloud forest from southwestern Colombia southward through Peru.

A Collared Inca (*Coeligena torquata*), an Andean species, flashes black and white in flight.

Tiny, common, and aggressive toward other hummers, the Copper-rumped Hummingbird (*Amazilia tobaci*) lives in Venezuela and on Trinidad and Tobago.

The Cuban Emerald (*Chlorostilbon ricordii*) inhabits its namesake archipelago as well as the islands of the northwestern Bahamas.

Despite being common in its namesake country, the Cuban Emerald (female shown here) has been found only a handful of times in nearby Florida.

The Ecuadorian Hillstar (*Oreotrochilus chimborazo*) is the alpine champion of hummingbirds, living at elevations from 11,000 to more than 17,000 feet on the volcanic peaks of the Andes.

The Fiery-throated Hummingbird (*Panterpe insignis*). These stunners live in the mountains of Costa Rica and Panama.

The Fork-tailed Woodnymph (*Thalurania furcata*) is widespread, its range covering most of northern and central South America.

The Glittering-throated Emerald (*Amazilia fimbriata*), another widespread South American hummingbird.

The Golden-breasted Puffleg (*Eriocnemis mosquera*)—so called for the puffy tufts of white feathers enshrouding its legs.

The Golden-tailed Sapphire (*Chrysuronia oenone*) is an ever-ready photographic subject in Colombia and Venezuela.

Gould's Jewelfront (aka Gould's Brilliant; *Heliodoxa aurescens*) ranges across much of the interior of northern South America.

One of the largest hummingbirds, the Great Sapphirewing (*Pterophanes cyanopterus*) inhabits the Andes in Bolivia, Colombia, Ecuador, and Peru.

The Green-breasted Mango (*Anthracothorax prevostii*), glittering metallically from head to tail, is common from Mexico through much of Central America.

The Green-crowned Brilliant (*Heliodoxa jacula*) is found from the uplands of Costa Rica to the Andes in Ecuador, in myriad disjunct populations.

The Green-crowned Woodnymph (*Thalurania colombica* subsp. *fannyae*) is found on the Pacific slope from Panama to Ecuador.

A male Green Thorntail (*Discosura conversii*) displays the tail feathers that inspired the common name.

The female Green Thorntail may not be as shiny as the male, but she is smartly attired nonetheless. These tiny hummers range from Costa Rica to western Ecuador.

A mountain species, the Lesser Violetear (*Colibri cyanotus*) ranges from Costa Rica south into the Andes down to Argentina as well as eastward into Venezuela.

The amazing Long-tailed Sylph (*Aglaiocercus kingii*), a hummingbird of the Andes.

Though lacking the brilliance and bannerlike tail of the male, the female Long-tailed Sylph is nonetheless a beautiful bird.

The Orange-throated Sunangel (*Heliangelus mavors*) is found only in the Andes of northwestern Venezuela and adjacent Colombia.

In its limited range in the montane forests of Colombia, Ecuador, and Peru, the Pink-throated Brilliant (*Heliodoxa gularis*) is increasingly imperiled by deforestation and is now ranked as near threatened.

The Purple-crowned Plovercrest (*Stephanoxis loddigesii*) is restricted to a relatively small region of southern Brazil and adjacent northeastern Argentina.

The male Purple-throated Mountain-gem (*Lampornis calolaemus*) is one of the more flamboyant hummingbirds of the Central American mountains.

Purple-throated Mountain-gems are strongly sexually dimorphic, as witnessed by the female, shown here.

The amazing Rufous-crested Coquette (*Lophornis delattrei*), a woodland species that occurs in disjunct populations in Panama, Colombia, Ecuador, Peru, and Bolivia.

The Rufous-gaped Hillstar (*Urochroa bougueri*) is a denizen of the northern Andes.

The Rufous-tailed Hummingbird (*Amazilia tzacatl*) is common from southern Mexico to the northern tier of South America.

The Saw-billed Hermit (*Ramphodon naevius*) is an understory species found only in southeastern Brazil. This is a male, as identified by the hooked beak.

Typical of the hermits, the Scale-throated Hermit (*Phaethornis eurynome*) is not particularly colorful. It resides in southern Brazil and adjacent Paraguay and northeastern Argentina.

The Scintillant Hummingbird (*Selasphorus scintilla*) is endemic to the Pacific slope of Costa Rica and Panama. If it looks familiar, that's because it's in the same genus as the Rufous and Allen's Hummingbirds.

The Shining Sunbeam (*Aglaeactis cupripennis*) is a robust hummingbird of the mountains, ranging through the Andes of Colombia and Peru.

Eastern Brazil's Sombre Hummingbird (*Aphantochroa cirro-chloris*) proves that not all hummers have bedazzling plumage.

The unmistakable Snowcap (*Microchera albocoronata*) of Central America can appear black, purplish, or bronze, but the white crown is diagnostic.

The stunning Red-billed Streamertail (*Trochilus polytmus*), Jamaica's national bird.

The spectacular Swallow-tailed Hummingbird (*Eupetomena macroura*), found throughout much of Brazil and surrounding countries, is one of the few species in which the male and female plumage is quite similar. Females are smaller and slightly less shiny.

The bizarre but aptly named Sword-billed Hummingbird (*Ensifera ensifera*) hails from the Andes. The male, shown here, is greener and more glittery than the female.

The Talamanca Hummingbird (*Eugenes spectabilis*), a colorful denizen of mountain ranges in Costa Rica and western Panama.

The Tufted Coquette (*Lophornis ornatus*) is an ornately plumed beauty from Trinidad and northeastern South America, including parts of Venezuela, Guyana, and Brazil.

The Tyrian Metaltail (*Metallura tyrianthina*) is widespread in the montane forests of Venezuela, Colombia, Ecuador, and extreme northern Peru.

Upon landing, the Velvet-purple Coronet (*Boissonneaua jardini*) often briefly extends its wings straight up, flashing cinnamon-colored underwings. It occurs on the Pacific slope of the Andes in Colombia and northern Ecudaor.

In ideal light, the Versicolored Emerald (*Amazilia versicolor*) displays subtly beautiful colors. This species is widespread in South America.

The vibrant Violet-capped Woodnymph (*Thalurania glaucopis*) ranges across southern Brazil and adjacent Paraguay, Argentina, and Uruguay.

The Violet-crowned Woodnymph (*Thalurania colombica*), among the most dazzling of hummingbirds, is widespread in Central America and northern South America.

The largest and one of the most conspicuous hummingbirds in Central America, the Violet Sabrewing (*Campylopterus hemileucurus*) ranges from southern Mexico to Panama.

The Violet-tailed Sylph (*Aglaiocercus coelestis*), found on the west slope of the Andes in Colombia and Ecuador, is understandably a favorite among hummingbird enthusiasts.

The female Violet-tailed Sylph is beautiful in her own right, although she lacks the bannerlike tail and shimmering colors of her counterpart.

The White-necked Jacobin (*Florisuga mellivora*) is widespread, ranging from southern Mexico to Bolivia and from coast to coast across northern South America.

Though not as showy as the male, the female White-necked Jacobin is distinctive nonetheless, with her conspicuously scaled throat.

The handsome White-throated Hummingbird (*Leucochloris albicollis*) is found in southeastern Brazil, northeastern Argentina, Paraguay, and Uruguay.

The gorgeous White-throated Mountain-gem (*Lampornis castaneoventris*) is divided into two subspecies; this is the male of subsp. *cinereicauda*.

The White-throated Mountain-gem (female shown here) is found only in the mountains of Costa Rica and Panama.

Named for the large white patch of undertail covert feathers, the White-vented Plumeleteer (*Chalybura buffonii*) is a large hummingbird that lives in parts of Panama, Venezuela, Colombia, and Ecuador.

Acknowledgments

Many different photographers contributed their amazing images to this book, not only making this work graphically rich but, in fact, making the entire project tenable. To an international cadre of outstanding photographers who are also hummingbird aficionados, I extend profound thanks: Nagi Aboulenein, Nick Athanas, Christine Baird, David Baker, Paul Baker, Tony Battiste, Andrea Belcher, Matt Betts, David Boyarski, Priscilla Burcher, Eduardo Castro, Tom Crabtree, Denise Dewire, Peter Edmonds, Matt Eldridge, Judy Royal Glenn, Eric Gofreed, Kim Guertin, Rosemary Harris, Gerald Hoekstra, Miroslaw Krol, Fabio N. Manfredini, Mike Martin, Samuel McCloud, Jessica McConahay, Andy Morffew, Verne Nelson, Roy Priest, Larry Reis, Yamil Saenz, Vic Schendel, Andreas Schmalz, Jeison MacGregor Solano, Philip Sorrentino, Bart Stegman, Stephen Vaughan, Roger Zachary, and Darin Ziegler. In addition, as the photo credits show, this book includes dozens of photos sourced through Creative Commons; I sincerely thank all these photographers for making their captivating work available for public and commercial use.

Anna's Hummingbird.

And I appreciate all the valuable input provided for the asking by hummingbird experts, birding enthusiasts, Audubon Society chapter members, and public facilities personnel. Finally, a heartfelt thanks to the intrepid Will McKay, Timber Press acquisitions editor, for navigating a path for allowing this book to come to fruition through a wonderful partnership with the invaluable *Birds & Blooms* magazine.

Useful Resources

BOOKS

Fogden, Michael, et al. 2014. *Hummingbirds: A Life-size Guide to Every Species*. New York: Harper Designs.

Howell, Steve N. G. 2004. *Hummingbirds of North America: The Photographic Guide*. Princeton, New Jersey: Princeton University Press.

Kavanagh, James, and Raymond Leung. 2017. *Hummingbirds: A Folding Pocket Guide to Familiar Species*. Dunedin, Florida: Waterford Press.

West, George C. 2015. *North American Hummingbirds: An Identification Guide*. Albuquerque: University of New Mexico Press.

Williamson, Sheri L. 2002. *Hummingbirds of North America*. Boston: Houghton Mifflin Harcourt.

ONLINE

Arizona Bird Committee
abc.azfo.org

Audubon Society
audubon.org

BirdNote
birdnote.org

Cornell Lab of Ornithology
birds.cornell.edu/home
birdsoftheworld.org
ebird.org

Hilton Pond Center for Piedmont Natural History
hiltonpond.org

Hummingbird Monitoring Network
hummonnet.org

Hummingbird Research, Inc.
hummingbirdresearch.net

Hummingbirds.net
hummingbirds.net

Hummingbird Society
hummingbirdsociety.org

I Found A Hummingbird
ifoundahummingbird.com

Operation RubyThroat
rubythroat.org

Texas Breeding Bird Atlas
txtbba.tamu.edu

Tropical Birding
tropicalbirding.com/north-america

UC Davis Hummingbird Health and Conservation Program
hummingbirds.vetmed.ucdavis.edu

USGS Bird Banding Laboratory
usgs.gov/centers/pwrc/science/bird-banding-laboratory?qt-science_center_objects=0#qt-science_center_objects

Western Hummingbird Partnership
westernhummingbird.org

Photo Credits

chaileefung0/Pixabay, pages 2–3, 89 (top).
Darin Ziegler, pages 6, 114.
Philip Sorrentino, page 10.
Mark Gunn/Flickr, page 14.
David Boyarski, pages 16 (top), 53, 65.
Fabio N. Manfredini, pages 17 (top), 208 (bottom), 209 (top right), 210 (bottom), 218 (top right), 220 (right), 221 (bottom), 224 (top right), 226 (bottom).
Bernard Spragg/Flickr, page 17 (bottom).
cuatrok77/Flickr, page 18 (top).
GregorySlobirdSmith/Flickr, page 18 (bottom).
Rick from Alabama/Flickr, page 19 (top).
ryanacandee/Flickr, pages 19 (bottom left), 214 (bottom), 216 (bottom).
Dr. Alexey Yakovlev/Flickr, page 19 (bottom right).
Diego Delso, page 20.
Andreas Schmalz, pages 21, 182.
CharmaineZoe's Marvelous Melange/Flickr, page 22.
Remko van Dokkum/Flickr, page 24 (left).
Mark Morgan/Flickr, page 24 (right).
Becky Matsubara/Flickr, pages 25 (top), 135 (top right, bottom right), 138 (top right), 218 (bottom left), 225 (top right).
Andy Morffew, pages 25 (bottom), 210 (top), 211 (bottom left), 217 (top left), 222 (bottom).
Eric Gofreed, page 26 (bottom).
Vic Schendel, page 29.
Anne Reeves/Flickr, pages 30, 101, 116.
Mike's Birds/Flickr, page 31.
Stephen Vaughan, pages 32, 130, 132, 135 (bottom left), 142 (middle, bottom right), 145 (top left), 145 (bottom right), 148 (bottom right), 151 (top right), 157 (middle left, bottom right), 163 (top left, bottom), 166 (top right, bottom right), 172, 175 (top, bottom), 178 (middle).
Tracie Hall/Flickr, pages 33, 135 (top left).
Bart Stegman, page 36 (left).
Gary Eslinger/USFWS, page 36 (right).
Mike Martin, page 37 (bottom).
David Mitchell/Flickr, page 41 (bottom).
Shannon Smith/Pixabay, page 42.
Tom Koerner/USFWS, pages 43, 188.
Eduardo Castro, page 47.
James Wainscoat/Unsplash, pages 48, 110.
Mike Tungate/Flickr, page 49.
Panegyrics of Granovetter/Flickr, pages 50, 214 (top left), 218 (top left).
Matt Betts, page 51.
Larry Lamsa/Flickr, page 52.
USGS, page 55.
Tom Shockey/Flickr, page 56.
USFWS, page 58.
JennyLeeSilver/Flickr, page 60.
likeaduck/Flickr, page 62 (top left).
parasolgarden.com, page 62 (top right, bottom).
Christine Baird, page 68.
Jessica McConahay, pages 74, 102, 104.
Rosemary Harris, pages 77, 90, 92.
Patrick Standish/Flickr, page 78.
Deb Nystrom/Flickr, pages 80, 89.
Oakley Originals/Flickr, page 81.
Lani Hudelson/Flickr, page 86 (left).
Maja Dumat/Flickr, pages 86 (right), 111 (bottom).
Elizabeth Nicodemus/Flickr, page 91.
Mark Levisay/Flickr, page 93.
Verne Nelson, page 94.
JKehoe/Flickr, page 95.
Wendy Cutler/Flickr, page 97.
Carol Norquist/Flickr, page 98.
Helena Jacoba/Flickr, page 99.
Minder Cheng/Flickr, page 100.
Katrina J. Houdek/Flickr, page 103.
Damien Pollet/Flickr, page 105.
Melissa McMasters/Flickr, pages 106, 119, 129, 196.
Samuel McCloud, page 107.
Renee Grayson/Flickr, pages 108, 118.
Paul Asman and Jill Lenoble/Flickr, page 109.
ChuckB/Flickr, page 111 (top).
Ed Dunens/Flickr, page 112.
Jonathan Hover/Flickr, page 113.
KM/Flickr, page 115.
Sound_Gene/Flickr, page 117.

Priscilla Burcher, page 120.
Akira Takiguchi/Flickr, page 121 (left).
sailn1/Flickr, page 121 (right).
Miroslaw Krol, page 124.
Wendell Smith/Flickr, page 125.
Larry Reis, page 126.
cultivar413/Flickr, page 127.
Jim Hammer/Flickr, page 128.
Rocky Raybell/Flickr, page 142 (top left).
Tony Battiste, pages 145 (top right), 148 (bottom left).
Brandon Trentler/Flickr, pages 145 (bottom left), 209 (bottom).
Paul Baker, page 148 (top left).
Tom Crabtree, page 148 (top right).
David Baker, page 148 (middle).
Denise Dewire, page 151 (middle).
Yamil Saenz, pages 151 (bottom left and right), 212 (bottom left).
dfaulder/Flickr, pages 154 (top right), 184 (top).
Shenandoah National Park/Flickr, page 154 (bottom left).
Judy Royal Glenn, pages 154 (bottom right), 190 (top).
Tom Grey, page 163 (top right).
Nagi Aboulenein, pages 166 (top left), 180 (left, middle).
D. Alexander Carrillo Mtz/Flickr, page 166 (middle left).
Alan Schmierer/Flickr, pages 166 (middle right, bottom left), 180 (right).
Roy Priest, page 169 (left).
Roger Zachary, page 169 (top).
Gerald Hoekstra, page 169 (bottom).
Gary Leavens/Flickr, page 175 (middle).
Shawn Taylor/Flickr, page 178 (left).
Missouri Department of Conservation, pages 184 (bottom), 194 (bottom).
Lisa Sheirer/Flickr, page 187.
State Botanical Garden of Georgia, page 190 (bottom).
USDA Forest Service, page 192.
Andrea Belcher, page 194 (top).
Andrew Cannizzaro/Flickr, page 197.
Matt Eldridge/Norman Bird Sanctuary, page 198.
Jonathan Cutrer/Flickr, page 199.
Kim Guertin/Audubon Vermont, page 200.
Brinton Museum, page 201.
Chris Charles/Unsplash, page 204.
F Delventhal/Flickr, page 206 (top).
Rolf Riethof/Flickr, pages 206 (bottom), 207, 215 (bottom), 217 (top right).
Dario Sanches/Wikipedia, pages 208 (top left), 222 (top right).
Ekaterina Chernetsova (Papchinskaya)/Flickr, pages 208 (top right), 212 (top right).
Nina Hale/Flickr, page 209 (top left).
Felipe Uribe/Flickr, page 211 (top left).
CU Boulder Alumni Association/Flickr, page 211 (top right).
Feroze Omardeen/Flickr, pages 211 (bottom right), 223 (top right).
Charles J. Sharp/Wikipedia, page 212 (top left).
Joseph C. Boone/Wikipedia, pages 212 (bottom right), 225 (bottom right).
Don Henise/Flickr, page 213 (top left).
Fernando Flores/Flickr, page 213 (top right).
Peter Edmonds, pages 213 (bottom), 223 (bottom), 224 (left), 225 (bottom left).
Nick Athanas, pages 214 (top right), 219 (left), 220 (top left), 222 (left).
Andy Reago & Chrissy McClarren/Flickr, pages 215 (left), 216 (top right), 221 (top right), 225 (top left).
Sam May/Flickr, pages 215 (top right), 216 (left), 219 (bottom), 226 (top left).
Andres Cuervo/Flickr, page 217 (bottom).
Don Faulkner/Flickr, pages 218 (bottom right), 223 (top left), 227 (top).
Alejandro Bayer Tamayo/Wikipedia, pages 219 (top right), 227 (bottom).
Matt MacGillivray/Flickr, page 220 (bottom).
Jeison MacGregor Solano, page 221 (top left).
dany13/Flickr, page 224 (bottom).
Kathy&sam/Wikipedia, page 226 (top right).
Daniel Dionne/Flickr, page 228.

All other photos are either in the public domain or by the author.

Index

N

O

P

R

S

John Shewey

About the Author

Lifelong birding enthusiast John Shewey is a veteran writer, editor, and professional outdoor photographer, with credits in *Birdwatching, Portland Monthly, Northwest Travel & Life,* and dozens of other magazines, and co-author of *Birds of the Pacific Northwest,* a Timber Press Field Guide. John has photographed birds from the mountains of Alaska to the jungles of Central America to the islands of the Caribbean, and his website chronicles many of these travels in rich photographic detail. Visit him at birdingoregon.com.